MODEL THEORY FOR MODAL LOGIC

SYNTHESE LIBRARY

STUDIES IN EPISTEMOLOGY,

LOGIC, METHODOLOGY, AND PHILOSOPHY OF SCIENCE

VOLUME 127

KENNETH A. BOWEN
Syracuse University

MODEL THEORY FOR MODAL LOGIC

Kripke Models for Modal Predicate Calculi

D. REIDEL PUBLISHING COMPANY
DORDRECHT : HOLLAND / BOSTON : U.S.A.
LONDON : ENGLAND

Library of Congress Cataloging in Publication Data

CIP

Bowen, Kenneth A.
Model theory for modal logic.

(Synthese library; v. 127)
Bibliography: p.
Includes indexes.
1. Modality (Logic) 2. Model theory.
3. Predicate calculus. I. Title. II. Title:
Kripke models for modal predicate calculi.
QA9.46.B68 511'.3 78-13182
ISBN 90-277-0929-7

Published by D. Reidel Publishing Company,
P.O. Box 17, Dordrecht, Holland

Sold and distributed in the U.S.A., Canada, and Mexico
by D. Reidel Publishing Company, Inc.
Lincoln Building, 160 Old Derby Street, Hingham,
Mass. 02043, U.S.A.

Printed in The Netherlands

To Johanna

TABLE OF CONTENTS

PREFACE

Modal considerations in logic appeared in the work of the ancients, notably Aristotle, and the mediaeval logicians, but like most work before the modern period, it was non-symbolic, and not particularly systematic in approach. According to Hughes and Cresswell (1968), the earliest symbolic approaches to modal logic were those of MacColl, originating about 1880. However, the first symbolic *and* systematic approach to the subject appears to be the work of Lewis beginning in 1912 and culminating in the book *Symbolic Logic* with Langford in 1932. Since then a plethora of modal systems have been proposed, though Lewis' systems S1–S5 and the system called T by Feys (1937) and M by von Wright (1951) stand out as having received considerable attention.

By far the early emphasis was on propositional systems of modal logic (with the notable exception of the work of R. Barcan Marcus (1946a), (1946b) and (1947)) and prior to the 1950's the formal semantic work done had been of a topological or algebraic cast. The work of Kripke beginning in (1959) provided a structural semantics for many of these systems which, when applied to systems containing predicates and quantifiers applied to individual variables, provides a semantics whose general appearance is similar to that which has been constructed for classical predicate logic. Related work was begun about the same time by Hintikka (1961) and Montague (1960). With the exception of some work of Gabbay (1972a) and Osswald (1969) most of this formal work (as opposed to related philosophical applications) has consisted in establishing a semantics for the system and giving proofs of completeness relative to this semantics (cf. Føllesdal (1965), (1968), van Fraasen (1969), Hintikka (1961), (1963), (1967), Kripke (1959), (1963a), (1963b), (1965), Lemmon (1966), Makinson (1966), Routley (1970), Schütte (1970), Segerberg (1971), and Thomason (1970)). In Bowen (1975) it was shown that a considerable portion of the model theory of classical predicate logic could be transferred to the domain of normal modal logics (cf. §1), usually without serious distortion. Our purpose here is not only to give a systematic presentation of these results and further extensions, but also to show that the restriction to normal modal systems can largely be removed. However, we will still adopt the converse of the

Barcan formula (cf. §1); in fact, in the natural formulation of the systems considered, it is a provable statement. In §§1–4 we formulate the systems and their semantics. Since much of the previous work is scattered or inaccessible (notably Lemmon and Scott (1966)) we give a detailed presentation of the proofs of correctness and completeness in §§3–4. Model theory proper begins in §5.

This book grew out of a series of lectures given at the Stephan Banach International Mathematical Center in Warsaw during the spring of 1973, during which I was on sabbatical leave from Syracuse University and was also supported by the Center. I would especially like to thank all the participants in Prof. Rasiowa's seminar for their patience in my lectures and their many corrections and suggestions. Ester Clark persevered through the typing of the original manuscript, and Ruth Turnpaugh typed the appendix. For their patience I am most grateful. Much of the work presented here was partially supported by ARPA grant number DAHCO4-72-C-0003.

§1 SYNTACTIC CONSIDERATIONS

We will adopt a good deal (but not all) of the notations and conventions of Shoenfield (1967) used for classical predicate logic. In particular, some of our metavariables are as follows: **u**, **v**, ... for expressions, **A**, **B**, **C**, ... for formulas, **a**, **b**, **c**, ... for terms, and **x**, **y**, **z**, ... for variables. We will use

(1.1) $\mathbf{A}_{\mathbf{x}_1 \ldots \mathbf{x}_n}[\mathbf{a}_1, \ldots, \mathbf{a}_n]$

to indicate the result of simultaneously substituting the terms $\mathbf{a}_1, \ldots, \mathbf{a}_n$ for the variables $\mathbf{x}_1, \ldots, \mathbf{x}_n$, respectively, where for $i = 1, \ldots, n$, $\mathbf{a}_i$ is free for $\mathbf{x}_i$. If the free variables of **A** are $\mathbf{x}_1, \ldots, \mathbf{x}_n$ in alphabetic order (as given below), then $\mathbf{A}[\mathbf{a}_1, \ldots, \mathbf{a}_n]$ is an abbreviation of (1.1). The basic symbols of all our modal languages are as follows:

variables	$x \quad y \quad z \quad x' \quad y' \quad z' \quad x'' \quad \ldots$
equality symbol	$=$
logical symbols	$\neg \quad \vee \quad \exists \quad \Diamond$

In addition, a given modal language ML may contain n-ary function and predicate symbols for various $n \geq 0$ (individual constants are treated as 0-ary function symbols); these are called *nonlogical symbols.* We use **f** and **p** as metavariable for function and predicate symbols, respectively. A modal language ML is specified when the set of its nonlogical symbols is specified. If ML is a given modal language, we write L for the underlying classical language obtained by excising the possibility operator $\Diamond$ from ML. The formation rules of ML are those of L (Shoenfield, 1967, pp. 14–15) together with the rule: if **u** is a formula, then $\Diamond\mathbf{u}$ is a formula. We use the usual abbreviations and definitions of $\wedge$, $\rightarrow$, $\leftrightarrow$, and $\forall$, together with $\Box\mathbf{A}$ for $\neg\Diamond\neg\mathbf{A}$.

All of the modal logics we will consider contain the machinery of classical first order predicate logic with identity, say as given by Shoenfield (1967), pp. 20–22. Note that in all the axiom and rule schemata listed there, the formulas **A**, **B**, and **C** may now contain occurrences of $\Diamond$ (and hence $\Box$). If **A** is a formula of ML which is an instance of a theorem-schema of classical predicate logic with identity, we will refer to **A** as a *classical theorem*.

The systems we will consider are obtained by adding various axiom and rule schemata to the classical logical base. We will give a master list of these, where the rules are given in the form

$$\mathbf{A}_1, \ldots, \mathbf{A}_n/\mathbf{B}, \quad \text{proviso}$$

which indicates that **B** can be inferred from $\mathbf{A}_1, \ldots, \mathbf{A}_n$ if the given proviso is met. If n is a non-negative integer and **s** is one of the symbols $\Diamond$ or $\Box$, $\mathbf{s}^n$ denotes a sequence consisting of exactly n occurrences of **s**.

A0. $\Box(\mathbf{A} \to \mathbf{B}) \to_{\blacksquare} \Box\mathbf{A} \to \Box\mathbf{B}$.
A1. $\Box\mathbf{A} \to \mathbf{A}$.
A2. $\Box\mathbf{A} \to \Diamond\mathbf{A}$.
A3. $\Box\mathbf{B} \to_{\blacksquare} \mathbf{A} \to \Box\Diamond\mathbf{A}$.
A4. $\Box(\mathbf{A} \to \mathbf{B}) \to_{\blacksquare} \Box(\Box\mathbf{A} \to \Box\mathbf{B})$.
A5. $\Box x = x \to \Box\Box x = x$.
A6. $\Box\mathbf{A} \to \Box\Box\mathbf{A}$.
A7. $\Box\mathbf{B} \to_{\blacksquare} \Diamond\mathbf{A} \to \Box\Diamond\mathbf{A}$.
A8. $\Box\mathbf{B} \to \Box(\Box\mathbf{A} \to \mathbf{A})$.
A9. $\Box\mathbf{B} \to_{\blacksquare} \mathbf{A} \to \Box\mathbf{A}$.
A10. $\neg\Box\Box\mathbf{A}$.
A11. $\Diamond\Box\mathbf{A}$.
A12. $\Diamond\mathbf{A}$.
A13. $(\Box\Diamond\Box\mathbf{A} \to \Box\mathbf{A})\&(\Box\mathbf{A} \to \mathbf{A})$.
A14. (m, n, p, q). $\Box B \to_{\blacksquare} \Diamond^m\Box^n\mathbf{A} \to \Box^p\Diamond^q\mathbf{A}$.

If **A** is one of the axioms listed above, $\Box\mathbf{A}$ indicates the schema obtained by prefixing $\Box$ before the schema of **A**.

R1. $\mathbf{A}, \quad \mathbf{A} \to \mathbf{B}/\mathbf{B}$.
R2. $\mathbf{A} \to \mathbf{B}/\Box\mathbf{A} \to \Box\mathbf{B}$.
R2′. $\Box(\mathbf{A} \to \mathbf{B})/\Box(\Box\mathbf{A} \to \Box\mathbf{B})$.
R3. $\mathbf{A}/\Box\mathbf{A}$.
R3$_t$. $/\Box\mathbf{A}$, provided **A** is a classical theorem.
R4. $\Box\mathbf{A}/\mathbf{A}$.

The different systems are then described by the various axiom and rule schemata which they adopt:

S0.5°: A0 + R3$_t$ S2°: A0 + $\Box$A0 + R2′ + R3$_t$ + R4
C2: A0 + R2 D2: C2 + A2
E2: C2 + A1 S2: A0 + A1 + $\Box$A0 + $\Box$A1 + R2′ + R3$_t$
S3: A1 + A4 + $\Box$A1 + $\Box$A4 + R3$_t$ I: A0 + R3

M: $\quad$ I + A1 $\quad$ B: M + A14(0, 0, 1, 1)($= \mathbf{A} \to \Box\Diamond\mathbf{A}$)
S4: $\quad$ M + A14(0, 1, 2, 0)($= \Box\mathbf{A} \to \Box\Box\mathbf{A}$)
S5: $\quad$ M + A14(1, 0, 1, 1)($= \Diamond\mathbf{A} \to \Box\Diamond\mathbf{A}$).

A *system* S of modal logic is said to be specified when its axiom and rule schemata have been specified, as in the examples above. These axioms and rules are known as *logical* axioms and rules. Thus S is essentially a 'schema of schemata' any of whose concrete realizations are determined by the choice of a particular modal language ML in which S is to be formulated. Figure 1 indicates the inclusion relationships between some of the systems specified above, where an arrow from one system to another indicates that the system at the foot of the arrow is properly contained in the system at the tip. (Specifically, $S \to S'$ indicates that all theorems of S are provable in S′; it does not imply that S′ is closed under the rules of S.) For more details on these systems as well as others, cf. Feys (1965), Hughes and Cresswell (1972), Lemmon (1957), (1966), and Zeman (1973).

DEFINITION 1.2. Let S be a specified modal system.

(i) S is a *Lemmon system* if S is obtained from C2 by adding some formulas as additional axioms (schemata may be used) or the rule R3 or both.

(ii) S is a *Feys system* if S is obtained from S2 by addition of axioms of the form $\Box\mathbf{A}$ or pairs of axioms of the form $\mathbf{A}$, $\Box\mathbf{A}$ or the rule R3 or both.

(iii) S is a *semi-normal* system if it is obtained from I by addition of new axioms.

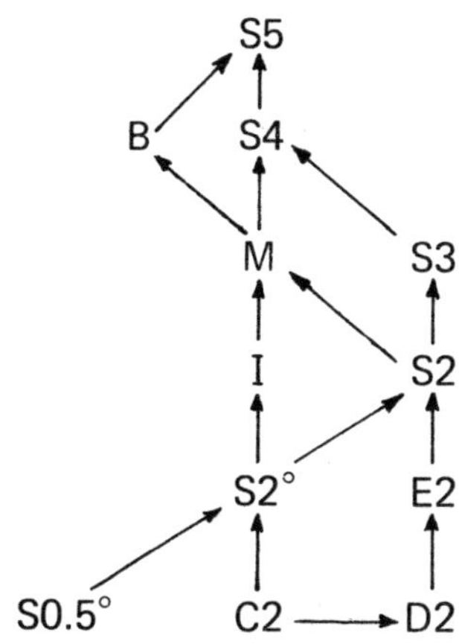

Fig. 1.

(iv) S is *normal* if S is semi-normal and contains A1. It is understood here that the formulas added are instances of modal, propositional schemata.

We will be particularly concerned with systems in which the axioms added are from the list A1–A14 or $\Box$A1–$\Box$A14. If S is a specified modal system and **A** is a formula of a modal language ML, we will write $\vdash^{S}\mathbf{A}$ to mean that there exists a proof of **A** in S, where as usual, a *proof of* **A** *in* S is a finite sequence of formulas the last of which is **A** such that each formula in the list is either an axiom of S or follows from earlier formulas in the list by one of the rules of inference of S.

Since all of our systems include classical predicate logic, one would expect that some standard theorems about that logic might extend to the present context, and indeed they do. In particular, if **A** is a replacement instance of a classical propositional tautology, then **A** is derivable in each of our systems. This is perhaps most easily seen by using the proof of the Tautology Theorem as given in Shoenfield (1967), the only modification necessary being the inclusion of formulas of the form $\Diamond\mathbf{A}$ among the *elementary formulas*. Moreover, in the terminology of Shoenfield (1967) (page references are to that book), the following all hold for each of the present systems:

(1.3) The ∀-Introduction and Generalization Rules (p. 31)
(1.4) The Substitution Rule and Theorem (pp. 31–32)
(1.5) The Distribution Rule and Closure Theorem (p. 32)
(1.6) The Theorem on Constants (p. 33)
(1.7) The Symmetry Theorem (p. 35)

For S containing the rule R2, the following theorem is proved by induction on the structure of **A**, while for S = S2° one uses Lemmon's (1957) procedure for showing that P2 contains S2 (cf. Hughes and Cresswell (1971), pp. 250–251). We omit the details.

THEOREM 1.8 (Substitutivity of Strict Equivalence). If S is a Lemmon or Feys system formulated in ML, if **A**, **A**′, **B**, and **B**′ are formulas of ML such that **A**′ is obtained by replacing zero or more occurrences of **B** in **A** by **B**′, then $\vdash^{S}\Box(\mathbf{B}\leftrightarrow\mathbf{B}')$ implies $\vdash^{S}\mathbf{A}\leftrightarrow\mathbf{A}'$.

THEOREM 1.9 (Substitutivity of Strict Equality). If S is a Lemmon or Feys system formulated in ML, if **A** and **A**′ are formulas of ML and **b** and **b**′ are terms of ML such that A′ is obtained from **A** by replacing zero or more occurrences of **b** by **b**′, then $\vdash^{S}\Box(\mathbf{b}=\mathbf{b}')$ implies that $\vdash^{S}\mathbf{A}\leftrightarrow\mathbf{A}'$.

Proof. Let $\mathbf{p}$ be a predicate symbol, and let $\mathbf{a}_1, \ldots, \mathbf{a}_n$ be terms, and let $\mathbf{a}'_i$ be the result of replacing zero or more occurrences of $\mathbf{b}$ in $\mathbf{a}_i$ by $\mathbf{b}'$ for $i = 1, \ldots, n$. Then $\mathbf{b} = \mathbf{b}' \rightarrow \mathbf{pa}_1 \ldots \mathbf{a}_n \rightarrow \mathbf{pa}'_1 \ldots \mathbf{a}'_n$ is a classical theorem and so by R3$_t$, A0, and modus ponens,

$$\vdash^{S} \Box(\mathbf{b} = \mathbf{b}') \rightarrow \Box(\mathbf{pa}_1 \ldots \mathbf{a}_n \rightarrow \mathbf{pa}'_1 \ldots \mathbf{a}'_n),$$

so if $\vdash^{S} \Box(\mathbf{b} = \mathbf{b}')$, then

$$\vdash^{S} \Box(\mathbf{pa}_1 \ldots \mathbf{a}_n \rightarrow \mathbf{pa}'_1 \ldots \mathbf{a}'_n).$$

Now Theorem 1.8. can be used to obtain the desired result.■

THEOREM 1.10 (Substitutivity of Equivalence). If S is a Lemmon system, $\mathbf{A}$, $\mathbf{A}'$, $\mathbf{B}$, and $\mathbf{B}'$ are formulas of ML such that $\mathbf{A}'$ is obtained from $\mathbf{A}$ by replacing one or more occurrences of $\mathbf{B}$ by $\mathbf{B}'$, then $\vdash^{S} \mathbf{B} \leftrightarrow \mathbf{B}'$ implies $\vdash^{S} \mathbf{A} \leftrightarrow \mathbf{A}'$.

Proof. By induction on the structure of $\mathbf{A}$. We omit the details.

The schema

$$(1.11) \qquad \forall \mathbf{x} \Box \mathbf{A} \rightarrow \Box \forall \mathbf{x} \mathbf{A}$$

is generally known as the *Barcan formula.* While it is not valid in any of our systems (cf. Kripke, 1963b), its converse is provable.

THEOREM 1.12. If S is any Lemmon or Feys system, we have $\vdash^{S} \Box \forall \mathbf{x} \mathbf{A} \rightarrow \forall \mathbf{x} \Box \mathbf{A}$.

Proof. In any system containing R2, the following is a proof:

$\forall \mathbf{x} \mathbf{A} \rightarrow \mathbf{A}$	Substitution Theorem
$\Box \forall \mathbf{x} \mathbf{A} \rightarrow \Box \mathbf{A}$	R2
$\Box \forall \mathbf{x} \mathbf{A} \rightarrow \forall \mathbf{x} \Box \mathbf{A}$	$\forall$-Introduction.

In S0.5°, the transition from the first to the second step is accomplished by R3$_t$, A0, and modus ponens. ■

If S is a Lemmon or Feys system, we say that an S-*theory* T has been specified if the modal language ML(T) in which T is to be formulated has been indicated and if some set of formulas, possibly empty and not including any logical axioms of S, has been indicated as the set of *non-logical axioms* of T. The *axioms* of T are made up of the logical axioms of S and the nonlogical axioms of T. If T is an S-theory and $\mathbf{A}$ is a formula of ML(T) then $\mathbf{A}$ is *provable in* T (*with respect to* S) (write $\vdash^{S}_{T}\mathbf{A}$) if there are nonlogical axioms $\mathbf{B}_1, \ldots, \mathbf{B}_n$ of T such that

$$\vdash^{S}\neg\mathbf{B}_1 \vee \cdots \vee \neg\mathbf{B}_n \vee \mathbf{A}.$$

If T is an S-theory and Γ is a set of formulas of ML(T), T(Γ) indicates the S-theory whose nonlogical axioms consist of the nonlogical axioms of T together with all formulas in Γ. In systems in which they occur, the rules R2, R2′, R3 and R3$_t$ will be referred to as *modal rules*. The other rules will be *nonmodal* rules.

LEMMA 1.13. (a) If S is a Lemmon system and $\vdash^{S}\Box(\mathbf{A}\to\mathbf{B})\to\Box(\Box\mathbf{A}\to\Box\mathbf{B})$, then $\vdash^{S}\Box\mathbf{B}\to\Box(\Box\mathbf{A}\to\Box\mathbf{B})$. (b) If S is a Feys system and $\vdash^{S}\Box(\Box(\mathbf{A}\to\mathbf{B})\to\Box(\Box\mathbf{A}\to\Box\mathbf{B}))$, then $\vdash^{S}\Box(\Box\mathbf{B}\to\Box(\Box\mathbf{A}\to\Box\mathbf{B}))$.

Proof. For (a), from $\vdash^{S}\mathbf{B}\to(\mathbf{A}\to\mathbf{B})$ we get $\vdash^{S}\Box\mathbf{B}\to\Box(\mathbf{A}\to\mathbf{B})$; then use the transitivity of implication. For (b), from $\vdash^{S}\mathbf{B}\to(\mathbf{A}\to\mathbf{B})$ we get $\vdash^{S}\Box(\mathbf{B}(\mathbf{A}\to\mathbf{B}))$ by R3$_t$, and then by R2′, $\vdash^{S}\Box(\Box\mathbf{B}\to\Box(\mathbf{A}\to\mathbf{B}))$. Now from $\vdash^{S}(\mathbf{C}\to\mathbf{D})\to_{\blacksquare}(\mathbf{D}\to\mathbf{E})\to_{\blacksquare}\mathbf{C}\to\mathbf{E}$ by R3$_t$, A0 and modus ponens, we get $\vdash^{S}\Box(\mathbf{C}\to\mathbf{D})\to\Box(\mathbf{D}\to\mathbf{E})\to\Box(\mathbf{C}\to\mathbf{E})$. Letting $\mathbf{C}$ be $\Box\mathbf{B}$, $\mathbf{D}$ be $\Box(\mathbf{A}\to\mathbf{B})$, and $\mathbf{E}$ be $\Box(\Box\mathbf{A}\to\Box\mathbf{B})$, the result now follows. ∎

LEMMA 1.14. Let S be any Lemmon system. Then for any n and m and any $\mathbf{A}_1, \ldots, \mathbf{A}_m$,

(i) $$\vdash^{S}\Box^n(\mathbf{A}_1 \wedge \ldots \wedge \mathbf{A}_m) \leftrightarrow \Box^n\mathbf{A}_1 \wedge \ldots \wedge \Box^n\mathbf{A}_m,$$

and

(ii) $$\vdash^{S}\Box^n\mathbf{A}_1 \vee \ldots \vee \Box^n\mathbf{A}_m \to \Box^n(\mathbf{A}_1 \vee \ldots \vee \mathbf{A}_m).$$

Proof. It will suffice to demonstrate the result for S = C2. For $n = 0$, both statements are trivialities, as it is also for $m = 1$. For $n = 1$ we proceed by induction on m, and for this it suffices to show that

$$\Box(\mathbf{B}\wedge\mathbf{C}) \leftrightarrow \Box\mathbf{B}\wedge\Box\mathbf{C} \quad \text{and} \quad \Box\mathbf{B}\vee\Box\mathbf{C}\to\Box(\mathbf{B}\vee\mathbf{C}).$$

For the first, since $\vdash^{C2}\mathbf{B}\wedge\mathbf{C}\to\mathbf{B}$ and $\vdash^{C2}\mathbf{B}\wedge\mathbf{C}\to\mathbf{C}$, then by R2, $\vdash^{C2}\Box(\mathbf{B}\wedge\mathbf{C})\to\Box\mathbf{B}$ and $\vdash^{C2}\Box(\mathbf{B}\wedge\mathbf{C})\to\Box\mathbf{C}$, so $\vdash^{C2}\Box(\mathbf{B}\wedge\mathbf{C})\to\Box\mathbf{B}\wedge\Box\mathbf{C}$. Conversely, we have $\vdash^{C2}\mathbf{A}\to(\mathbf{B}\to\mathbf{A}\wedge\mathbf{B})$, so by R2,

$$\vdash^{C2}\Box\mathbf{A}\to\Box(\mathbf{B}\to\mathbf{A}\wedge\mathbf{B}).$$

But by A0,

$$\vdash^{C2}\Box(\mathbf{B}\to\mathbf{A}\wedge\mathbf{B})\to_{\blacksquare}\Box\mathbf{B}\to\Box(\mathbf{A}\wedge\mathbf{B}).$$

so

$$\vdash^{C2} \Box \mathbf{A} \rightarrow \centerdot\, \Box \mathbf{B} \rightarrow \Box(\mathbf{A} \wedge \mathbf{B}),$$

and hence $\vdash^{C2} \Box \mathbf{A} \wedge \Box \mathbf{B} \rightarrow \Box(\mathbf{A} \wedge \mathbf{B})$. For the second needed result, we first have $\vdash^{C2} \mathbf{B} \rightarrow \mathbf{B} \vee \mathbf{C}$ and $\vdash^{C2} \mathbf{C} \rightarrow \mathbf{B} \vee \mathbf{C}$, so by R2,

$$\vdash^{C2} \Box \mathbf{B} \rightarrow \Box(\mathbf{B} \vee \mathbf{C}) \quad \text{and} \quad \vdash^{C2} \Box \mathbf{C} \rightarrow \Box(\mathbf{B} \vee \mathbf{C}),$$

so $\vdash^{C2} \Box \mathbf{B} \vee \Box \mathbf{C} \rightarrow \Box(\mathbf{B} \vee \mathbf{C})$. The induction steps for both n and m are now simple. ■

§2. MODAL STRUCTURES AND MORPHISMS*

A *modal structure base* or simply *base* is a quadruple $\langle \mathbf{K}, \mathbf{R}, \mathbf{O}, \mathbf{N} \rangle$, where $\mathbf{K}$ is a non-empty set, $\mathbf{O} \in \mathbf{K}$, $\mathbf{R}$ is a binary relation on $\mathbf{K}$, and $\mathbf{N} \subseteq \mathbf{K}$. We will usually write $\mathbf{Q}$ for $\mathbf{K} - \mathbf{N}$. The elements of $\mathbf{K}$ are usually referred to as *possible worlds*, $\mathbf{O}$ as the *origin* or *immediate situation*, $\mathbf{R}$ as the *accessibility relation* between the worlds of $\mathbf{K}$, $\mathbf{N}$ as the collection of *normal* worlds, and $\mathbf{Q}$ as the collection of *queer* worlds (cf. Kripke, 1963, 1965 and Lemmon, 1966).

DEFINITION 2.1. Let ML be a fixed modal language. A *modal structure* $\mathfrak{A}$ *for* ML consists of the following:

(i) a base $\langle \mathbf{K}, \mathbf{R}, \mathbf{O}, \mathbf{N} \rangle$.

(ii) for each $k \in \mathbf{K}$, a classical structure $\mathscr{A}_k = \langle A_k, \equiv_k, \mathbf{f}_k, \ldots, \mathbf{p}_k, \ldots \rangle$ for the underlying classical language L (cf. Shoenfield, 1967, p. 18) such that if $k, k' \in \mathbf{K}$ and $k \, \mathbf{R} \, k'$, then $A_k \subseteq A_{k'}$, where $\equiv_k$ is an equivalence relation on A_k which is a congruence relation for each $\mathbf{f}_k$ and $\mathbf{p}_k$; the relation $\equiv_k$ is used to interpret the equality symbol of ML*; this is a departure from the conventions of Shoenfield (1967), but cf. Kreisel and Krivine (1967, Chapter 3), and Robinson (1956), (1965), (1969).

We will usually write $\mathfrak{A} = \langle \mathscr{A}_k, \mathbf{K}, \mathbf{R}, \mathbf{O}, \mathbf{N} \rangle$ and we define the *base of* $\mathfrak{A}$ to be $b(\mathfrak{A}) = \langle \mathbf{K}, \mathbf{R}, \mathbf{O}, \mathbf{N} \rangle$.

DEFINITION 2.2. The language LB is the classical first order language with equality specified as follows:

variables:	$w \quad w' \quad w'' \quad \ldots$
equality:	$=$
unary predicate symbol:	N*
binary predicate symbol:	R*
individual constant:	O*
logical symbols:	$\neg \quad \vee \quad \exists$

* To some extent, our model theory interprets the equality symbol as 'equivalence' in the sense of Carnap (1947). More precisely, we ought to regard the entities inhabiting the universe $|\mathscr{A}_k|$ of the world $\mathscr{A}_k$ as intensional objects and take the relation $\equiv_k$ to be that of *extensional* identity between these intensional objects.

Thus for any modal structure $\mathfrak{A}$, $b(\mathfrak{A})$ is a classical structpre for the language LB. Note that for any modal language ML, the variables of ML and LB form disjoint sets. We will assume that the particular symbols R*, N*, and O* never occur in any modal language ML.

DEFINITION 2.3. Let Γ be a set of sentences of LB. A modal structure $\mathfrak{A}$ is said to be a *Γ-structure* if $b(\mathfrak{A})$ is a model of Γ. If ML is a modal language, we will write St(ML,Γ) for the class of all modal structures for ML which are Γ-structures.

DEFINITION 2.4. Let $\mathfrak{A} = \langle \mathscr{A}_k, \mathbf{K}, \mathbf{R}, \mathbf{O}, \mathbf{N} \rangle$ be a modal structure for the modal language ML. The *universe* of $\mathfrak{A}$ is defined to be

$$\mathrm{U}(\mathfrak{A}) = \cup \{ |\mathscr{A}_k| : k \in \mathbf{K} \},$$

where $|\mathscr{A}_k|$ is the universe of the classical structure $\mathscr{A}_k$. The *skeleton* of $\mathfrak{A}$ is defined to be

$$\mathrm{sk}(\mathfrak{A}) = \mathbf{K} \cup \mathrm{U}(\mathfrak{A}).$$

We will agree once and for all that for any modal structure $\mathfrak{A} = \langle \mathscr{A}_k, \mathbf{K}, \mathbf{R}, \mathbf{O}, \mathbf{N} \rangle$, $\mathbf{K} \cap \mathrm{U}(\mathfrak{A}) = \emptyset$.

DEFINITION 2.5. Let $\mathfrak{A}$ be a modal structure for the modal language ML. An *assignment in* $\mathfrak{A}$ is a map v from the set of variables of ML to $\mathrm{U}(\mathfrak{A})$.

DEFINITION 2.6. Let $\mathfrak{A}$ be a modal structure for the modal language ML, let $\mathbf{a}$ be a term of ML, let v be an assignment in $\mathfrak{A}$, and let $k \in \mathbf{K}$. We define the *denotation of* $\mathbf{a}$ *in* $\mathfrak{A}$ *at k under v*, $\mathbf{a}^{\mathfrak{A},k}[v]$, by recursion on the length of $\mathbf{a}$ as follows:*

(i) $\mathbf{x}^{\mathfrak{A},k}[v] = v(\mathbf{x})$.

$$\text{(ii)}\ (\mathbf{fa}_1 \ldots \mathbf{a}_n)^{\mathfrak{A},k}[v] = \begin{cases} \mathbf{f}_k(\mathbf{a}_1^{\mathfrak{A},k}[v], \ldots, \mathbf{a}_n^{\mathfrak{A},k}[v]), & \text{if} \\ \mathbf{a}_i^{\mathfrak{A},k}[v] \in |\mathscr{A}_k| & \text{for } i = 1, \ldots; \\ \mathbf{a}_j^{\mathfrak{A},k}[v], & \text{where } j \text{ is minimal} \\ \text{with } \mathbf{a}_j^{\mathfrak{A},k}[v] \notin |\mathscr{A}_k|, & \text{otherwise.} \end{cases}$$

* Note that variables are being treated as rigid designators in the sense of Kripke (1972), while the fact that individual constants are 0-ary function symbols admits the possibility that no closed terms might be rigid.

DEFINITION 2.7. Let $\mathfrak{A}$ be a modal structure for the modal language ML, let **A** be a formula of ML, let v be an assignment in $\mathfrak{A}$, and let $k \in \mathbf{K}$. The relation $\mathfrak{A} \models_k \mathbf{A}[v]$ is defined by recursion on the length of **A** as follows:

(i) $\mathfrak{A} \models_k \mathbf{a} = \mathbf{b}[v]$ iff $\mathbf{a}^{\mathfrak{A},k}[v] \equiv_k \mathbf{b}^{\mathfrak{A},k}[v]$.

(ii) $\mathfrak{A} \models_k \mathbf{p}\mathbf{a}_1 \ldots \mathbf{a}_n[v]$ iff $\mathbf{p}_k(\mathbf{a}_1^{\mathfrak{A},k}[v], \ldots, a_n^{\mathfrak{A},k}[v])$.

(iii) $\mathfrak{A} \models_k \mathbf{A} \vee \mathbf{B}[v]$ iff $\mathfrak{A} \models_k \mathbf{A}[v]$ or $\mathfrak{A} \models_k \mathbf{B}[v]$.

(iv) $\mathfrak{A} \vdash_k \neg \mathbf{A}[v]$ iff not $- \mathfrak{A} \models_k \mathbf{A}[v]$.

(v) $\mathfrak{A} \models_k \exists \mathbf{x} \mathbf{A}[v]$ iff for some $a \in |\mathscr{A}_k|$, $\mathfrak{A} \models_k \mathbf{A}[v(\substack{\mathbf{x} \\ a})]$, where $v(\substack{\mathbf{x} \\ a})$ is that assignment μ in $\mathfrak{A}$ such that $\mu(\mathbf{x}) = a$ and $\mu(\mathbf{y}) = v(\mathbf{y})$ for **y** distinct from **x**.

(vi) $\mathfrak{A} \models_k \Diamond \mathbf{A}[v]$ iff $k \in \mathbf{K} - \mathbf{N}$ or for some $k' \in \mathbf{K}$ with $k \mathbf{R} k'$, $\mathfrak{A} \models_{k'} \mathbf{A}[v]$.

We will write $\mathfrak{A} \models \mathbf{A}[v]$ if $\mathfrak{A} \models_{\mathbf{0}} \mathbf{A}[v]$. Clearly if **A** is a sentence then $\mathfrak{A} \models_k \mathbf{A}[v]$ is independent of the choice of v and so we write $\mathfrak{A} \models_k \mathbf{A}$ to mean that $\mathfrak{A} \models_k \mathbf{A}[v]$ for some (any) choice of v. Similarly for $\mathfrak{A} \models \mathbf{A}$. We say that a formula **A** is *valid in* $\mathfrak{A}$ if $\mathfrak{A} \models \mathbf{A}[v]$ for all assignments v in $\mathfrak{A}$ such that for all free variables **x** of **A**, $v(\mathbf{x}) \in |\mathscr{A}_\mathbf{0}|$ (we say that v is *local for* **A**;* more generally, if $v(x) \in |\mathscr{A}_k|$ for all free **x** in **A**, v is *local for* **A** *at* k). If T is an S-theory formulated in the modal language ML, an S-structure $\mathfrak{A}$ for ML is said to be an S-*model* of T if every axiom of T is valid in $\mathfrak{A}$. Two modal structures $\mathfrak{A}$, $\mathfrak{B}$ for ML are *elementarily equivalent* ($\mathfrak{A} \equiv \mathfrak{B}$) iff for all sentences **A** of ML, $\mathfrak{A} \models \mathbf{A}$ iff $\mathfrak{B} \models \mathbf{A}$.

DEFINITION 2.8. Let S be a modal system formulated in the modal language ML, and let Γ be a set of sentences of LB.

(i) S is *valid with respect to* Γ if for any formula **A** of ML and any structure $\mathfrak{A} \in \mathrm{St}(\mathrm{ML}, \Gamma)$, if $\vdash^{\mathrm{S}} \mathbf{A}$, then **A** is valid in $\mathfrak{A}$.

(ii) S is *complete with respect to* Γ if for any formula **A** of ML, if **A** is not derivable in S, then there exists a structure $\mathfrak{A} \in \mathrm{St}(\mathrm{ML}, \Gamma)$ and an assignment v in $\mathfrak{A}$ such that $\mathfrak{A} \models \neg \mathbf{A}[v]$ and v is local for **A**.

(iii) Γ is *characteristic for* S if S is both valid and complete with respect to Γ.

* Our semantic treatment of individuals and predicates is closely related to that of the system **Q3** of Thomason (1970); however, our restriction to local assignments in the definition of validity accounts for our syntactic differences from **Q3**. This restriction is natural, since in light of (2.7v) it preserves the classical fact that a formula is valid in a structure iff its universal closure is valid.

In §§3, 4 we will see that quite a few of the modal systems introduced in the literature possess characteristic sets Γ. Here we will describe some sentences of LB which will be used to construct those characteristic sets.

$\mathrm{Ref_N}$	$\forall w[N^*(w) \to w\mathrm{R}^*w]$.
$\mathrm{Sym_N}$	$\forall ww'[\mathrm{N}^*(w) \,\&\, \mathrm{N}^*(w') \,\&\, w\mathrm{R}^*w' \to w'\mathrm{R}^*w]$.
$\mathrm{Tran_N}$	$\forall w_1w_2w_3[\mathrm{N}^*(w_1) \,\&\, \mathrm{N}^*(w_2) \,\&\, w_1\mathrm{R}^*w_2 \,\&\, w_2\mathrm{R}^*w_3 \to w_1\mathrm{R}^*w_3]$.
$\mathrm{Cl_N}$	$\forall ww'[\mathrm{N}^*(w) \,\&\, w\mathrm{R}^*w' \to \mathrm{N}^*(w')]$.
Norm	$\forall w\mathrm{N}^*(w)$.
$\mathbf{x}_1\mathrm{R}^{*s}\mathbf{x}_2, s > 1$	$\exists \mathbf{y}_1 \ldots \exists \mathbf{y}_{s-1}[\mathbf{x}_1\mathrm{R}^*\mathbf{y}_1 \,\&\, \mathbf{y}_1\mathrm{R}^*\mathbf{y}_2 \,\&\, \ldots \,\&\, \mathbf{y}_{s-1}\mathrm{R}^*\mathbf{x}_2]$.
$x_1\mathrm{R}^{*1}x_2$	$x_1\mathrm{R}^*x_2$.
$x_1\mathrm{R}^{*0}x_2$	$x_1 = x_2$.
CD_1	$\mathrm{Ref_N}$.
CD_2	$\forall w[\mathrm{N}^*(w) \to \exists w'[w\mathrm{R}^*w']]$.
CD_3	$\mathrm{Sym_N}$.
CD_4	$\mathrm{Tran_N}$.
CD_5	$\mathrm{Cl_N}$.
CD_6	$\mathrm{Tran_N} \,\&\, \mathrm{Cl_N}$.
CD_7	$\forall w_1w_2w_3[\mathrm{N}^*(w_1) \,\&\, w_1\mathrm{R}^*w_2 \,\&\, w_1\mathrm{R}^*w_3 \to w_2\mathrm{R}^*w_3]$.
CD_8	$\forall w[\exists w'[\mathrm{N}^*(w') \,\&\, w'\mathrm{R}^*w] \to w\mathrm{R}^*w]$.
CD_9	$\forall ww'[\mathrm{N}^*(w) \,\&\, w\mathrm{R}^*w' \to w = w']$.
CD_{10}	$\forall w'[\mathrm{N}^*(w') \to \exists w[w'\mathrm{R}^*w \,\&\, \neg\, \mathrm{N}^*(w)]]$.
CD_{11}	$\forall w'[\mathrm{N}^*(w') \to \exists w[w'\mathrm{R}^*w \,\&\, \mathrm{N}^*(w) \,\&\, \neg \exists w''[w\mathrm{R}^*w'']]$.
CD_{12}	$\forall w \neg\, \mathrm{N}^*(w)$.
CD_{13}	$\mathrm{Sym_N} \,\&\, \mathrm{Ref_N}$.
$\mathrm{CD}_{14}(m, n, p, q)$	$\forall w_1w_2w_3[\mathrm{N}^*(w_1) \,\&\, w_1\mathrm{R}^{*m}w_2 \,\&\, w_1\mathrm{R}^{*p}w_3 \to \exists w_4[w_2\mathrm{R}^{*n}w_4 \,\&\, w_3\mathrm{R}^{*q}w_4]]$.

DEFINITION 2.9. Let S be a Lemmon system obtained by adding to C2 either R3 or some of A1–A14 or both. The set Γ_S is defined as follows:

(i) Γ_{C2} is empty.
(ii) If A_i is an axiom of S, then CD_i is in Γ_S.
(iii) If R3 is a rule of S, then Norm belongs to Γ_S.

DEFINITION 2.10. Let S be a Feys system obtained by adding to S2 either R3 or some of □A1–□A7, □A13, □A15, or both. Then the set Γ_S is defined as follows:

(i) Γ_{S2° consists of N*(O*).
(ii) If $\Box A_i$ is an axiom of S for $i = 1, \ldots, 7, 13$, then CD_i in in Γ_S.
(iii) If $\Box$A15 is an axiom of S, then CD_{10} is in Γ_S.
(iv) If R3 is a rule of S, then Norm belongs to Γ_S.

If S is one of the systems just considered and $\mathfrak{A}$ is a Γ_S-structure, we will usually refer to $\mathfrak{A}$ as an S-*structure*. Thus we will speak of C2-structures, S4-structures, etc. We will see in §§3–4 that for each of the specific Lemmon or Feys system S mentioned above, the set Γ_S is characteristic for S.

DEFINITION 2.11. Let $\mathfrak{A} = \langle \mathscr{A}_k, \mathbf{K}, \mathbf{R}, \mathbf{O}, \mathbf{N} \rangle$ and $\mathfrak{B} = \langle \mathscr{B}_l, \mathbf{L}, \mathbf{S}, \mathbf{P}, \mathbf{M} \rangle$ be modal structures for the modal language ML. A *protomorphism from* $\mathfrak{A}$ *to* $\mathfrak{B}$ is a map m: sk$(\mathfrak{A}) \to$ sk$(\mathfrak{B})$ such that:

(i) m maps $\mathbf{K}$ into $\mathbf{L}$ and $m(\mathbf{O}) = \mathbf{P}$.
(ii) $k \in \mathbf{N} \,\&\, k' \in \mathbf{K} \,\&\, k\mathbf{R}k' \to m(k)\mathbf{S}m(k')$.
(iii) $k \in \mathbf{K} - \mathbf{N} \to m(k) \in \mathbf{L} - \mathbf{M}$.
(iv) for each $k \in \mathbf{K}, a \in |\mathscr{A}_k| \to m(a) \in |\mathscr{B}_{m(k)}|$.
(v) for each $k \in \mathbf{K}$ and all $a, a' \in |\mathscr{A}_k|$, $a \equiv_k a' \leftrightarrow m(a) \equiv_{m(k)} m(a')$.

DEFINITION 2.12. A *monomorphism from* $\mathfrak{A}$ *to* $\mathfrak{B}$ is a protomorphism from $\mathfrak{A}$ to $\mathfrak{B}$ such that:

(i) for each n-ary function symbol $\mathbf{f}$ of ML, each $k \in \mathbf{K}$, and all $a_1, \ldots, a_n \in |\mathscr{A}_k|$, $m(\mathbf{f}_k(a_1, \ldots, a_n)) \equiv_{m(k)} \mathbf{f}_{m(k)}(m(a_1), \ldots, m(a_n))$;
(ii) for each n-ary predicate symbol $\mathbf{p}$ of ML (including $=$), each $k \in \mathbf{K}$, and all $a_1, \ldots, a_n \in |\mathscr{A}_k|$, $\mathbf{p}_k(a_1, \ldots, a_n) \to \mathbf{p}_{m(k)}(m(a_1), \ldots, m(a_n))$.
(iii) for each $k \in \mathbf{K}$ and all $a, a' \in |\mathscr{A}_k|$, $a \not\equiv_k a'$ implies $m(a) \not\equiv_{m(k)} m(a')$.

We will write $m: \mathfrak{A} \to \mathfrak{B}$ to indicate that m is a monomorphism from $\mathfrak{A}$ to $\mathfrak{B}$. Note that such an m is not necessarily one-one on the set of worlds $\mathbf{K}$. However, though m may collapse some worlds in $\mathbf{K}$, it always homomorphically preserves the structure of $\mathbf{R}$ and takes normal worlds onto normal worlds and queer worlds onto queer worlds. Given $k \in \mathbf{K}$, m may also fail to be one-one on $|\mathscr{A}_k|$. However, given $a, a' \in |\mathscr{A}_k|$, we can distinguish a and a' in $\mathscr{A}_k$ with $\equiv_k$ iff we can distinguish $m(a)$ and $m(a')$ in $\mathscr{B}_{m(k)}$ with $\equiv_{m(k)}$. Thus in fact, m induces a one-one map from the equivalence classes of $|\mathscr{A}_k|$ under $\equiv_k$ into $|\mathscr{B}_{m(k)}|$. If m is in fact one-one on $\mathbf{K}$, we say that m is a *strong monomorphism* and write $m: \mathfrak{A} \xrightarrow{S} \mathfrak{B}$. Finally, m is *faithful* if for all $k, k' \in \mathbf{K}$, $k\mathbf{R}k'$ iff $m(k)\mathbf{S}m(k')$.

DEFINITION 2.13. Let $\mathfrak{A}$ and $\mathfrak{B}$ be modal structures for ML and let $m: \mathfrak{A} \to \mathfrak{B}$. We say that m is an *elementary embedding of* $\mathfrak{A}$ *in* $\mathfrak{B}$ if for all formulas $\mathbf{A}$ of ML, all $k \in \mathbf{K}$, and all assignments v in $\mathfrak{A}$ which are local for $\mathbf{A}$ at k,

$$(2.14) \qquad \mathfrak{A} \models_k \mathbf{A}[v] \quad \text{iff} \quad \mathfrak{B} \models_{m(k)} \mathbf{A}[m \circ v].$$

In this case, we write $m: \mathfrak{A} \xrightarrow{\equiv} \mathfrak{B}$. If, in addition m is strong, we write $m: \mathfrak{A} \xrightarrow{S\equiv} \mathfrak{B}$. Finally, if $k \in \mathbf{K}$, the *truncation of* $\mathfrak{A}$ *at* k, $\mathrm{trc}_k(\mathfrak{A})$, is the structure $\langle \mathscr{B}_l, \mathbf{L}, \mathbf{S}, \mathbf{P}, \mathbf{M} \rangle$ where $\mathbf{L} = \{k\} \cup \{k' \in \mathbf{K}: k\mathbf{R}^{\infty}k'\}$, $\mathbf{S} = \mathbf{R} \upharpoonright (\mathbf{L} \times \mathbf{L})$, $\mathbf{P} = k$, $\mathbf{M} = \mathbf{N} \cap \mathbf{L}$, and for $l \in \mathbf{L}$, $\mathscr{B}_l = \mathscr{A}_l$, where $\mathbf{R}^{\infty}$ is the ancestral of $\mathbf{R}$.

THEOREM 2.15. Let $\mathfrak{A} = \langle \mathscr{A}_k, \mathbf{K}, \mathbf{R}, \mathbf{O}, \mathbf{N} \rangle$ and let $\mathfrak{A}\# = \langle \mathscr{A}_k, \mathbf{K}, \mathrm{R}\#, \mathbf{O}, \mathbf{N} \rangle$, where $\mathbf{R}\#$ is obtained from $\mathbf{R}$ by deleting all pairs $\langle k, k' \rangle$ such that $k \in \mathbf{K} - \mathbf{N} = \mathbf{Q}$. If id is the identity map on $\mathrm{sk}(\mathfrak{A}\#) = \mathrm{sk}(\mathfrak{A})$, then $\mathrm{id}: \mathfrak{A}\# \xrightarrow{S\equiv} \mathfrak{A}$.

Proof. In this case, (2.14) reduces to

$$(2.16) \qquad \mathfrak{A} \models_k \mathbf{A}[v] \quad \text{iff} \quad \mathfrak{A}\# \models_k \mathbf{A}[v].$$

We verify this by induction on the length of $\mathbf{A}$. If $\mathbf{A}$ is atomic or has $\neg$, $\vee$, or $\exists$ as its principal logical symbol, the procedure is simple. Suppose that $\mathbf{A}$ is $\diamond\mathbf{B}$. If $k \in \mathbf{Q}$, both sides of (2.16) hold by (2.7vi). So suppose that $k \in \mathbf{N}$. Then using (2.7vi) and the induction hypothesis

$$\begin{aligned}\mathfrak{A} \models_k \diamond\mathbf{B}[v] \quad &\text{iff} \quad \vee k' \in \mathbf{K}[k\mathbf{R}k' \,\&\, \mathfrak{A} \models_{k'} \mathbf{B}[v]], \\ &\text{iff} \quad \vee k' \in \mathbf{K}[k\mathbf{R}\#k' \,\&\, \mathfrak{A}\# \models_{k'} \mathbf{B}[v]], \\ &\text{iff} \quad \mathfrak{A}\# \models_k \diamond\mathbf{B}[v],\end{aligned}$$

noting that since $k \in \mathbf{N}$, $k\mathbf{R}k'$ iff $k\mathbf{R}\#k'$. ■

In view of Theorem 2.15, we will henceforth assume that all our structures $\mathfrak{A} = \langle \mathscr{A}_k, \mathbf{K}, \mathbf{R}, \mathbf{O}, \mathbf{N} \rangle$ are such that $\mathbf{R} \cap (\mathbf{Q} \times \mathbf{K}) = \emptyset$; i.e.,

$$b(\mathfrak{A}) \models \forall w[\neg \mathrm{N}^*(w) \rightarrow \neg \exists w' w\mathrm{R}^* w'].$$

Thus if k is a queer world, no world is accessible from k, not even k itself.

Given a base $\mathscr{B} = \langle \mathbf{K}, \mathbf{R}, \mathbf{O}, \mathbf{N} \rangle$, we write, as above, $k\mathbf{R}^n k'$ to mean that for $n = 1$, $k\mathbf{R}k'$, and for $n > 1$, there are $k_1, \ldots, k_{n-1} \in \mathbf{K}$ such that $k\mathbf{R}k_1$, $k_i\mathbf{R}k_{i+1}$ for $i = 1, \ldots, n-2$, and $k_{n-1}\mathbf{R}k'$. (We are counting the number of 'links' between k and k'.) We say that $\mathscr{B}$ is *weakly connected* if for all $k \in \mathbf{K}$, $k \neq \mathbf{O}$, there is an n so that $\mathbf{O}\mathbf{R}^n k$. We define the *heart of* $\mathscr{B}$, $h(\mathscr{B})$, to be $\mathscr{B} \upharpoonright h(\mathbf{K})$, where

$$h(\mathbf{K}) = \{\mathbf{O}\} \cup \{k \in \mathbf{K}: \quad \text{for some} \quad n \geq 1, \mathbf{O}\mathbf{R}^n k\}.$$

If $\mathfrak{A} = \langle \mathscr{A}_k, \mathbf{K}, \mathbf{R}, \mathbf{O}, \mathbf{N} \rangle$ is a modal structure and $\mathbf{O} \in \mathrm{L} \subseteq \mathbf{K}$, we define $\mathfrak{A} \upharpoonright \mathbf{L}$ to be the modal structure $\langle \mathscr{B}_l, \mathbf{L}, \mathbf{S}, \mathbf{P}, \mathbf{M} \rangle$ where $\mathbf{S} = \mathbf{R} \upharpoonright \mathbf{L}$, $\mathbf{P} = \mathbf{O}$, $\mathbf{M} = \mathbf{N} \cap \mathbf{L}$, and for $l \in \mathbf{L}$, $\mathscr{B}_l = \mathscr{A}_l$. Then we define the *heart of* $\mathfrak{A}$, $h(\mathfrak{A})$, to be $\mathfrak{A} \upharpoonright h(\mathbf{K})$.

LEMMA 2.17. If S is one of the modal systems specified in §2 and $\mathfrak{A}$ is an S-structure, then $h(\mathfrak{A})$ is also an S-structure.

Proof. All of the conditions CD1–CD14 and Norm are either universal sentences or of the form $\forall w_1 \ldots w_n \exists w[w_i \mathrm{R}^n w \,\&\, \mathbf{D}]$, where $\mathbf{D}$ is open. It follows from the definition of $h(b(\mathfrak{A}))$ and the Łos–Tarski Theorem (cf. Shoenfield, 1967, p. 76) that if any of the these sentences holds in $b(\mathfrak{A})$, it also holds in $h(b(\mathfrak{A}))$. ■

LEMMA 2.18. If $\mathfrak{A}$ is any modal structure, then $h(\mathfrak{A})$ is an elementary substructure of $\mathfrak{A}$.

Proof. Let v be any assignment in $h(\mathfrak{A})$ and let $k \in h(\mathbf{K})$. We show that $h(\mathfrak{A}) \models_k \mathbf{A}[v]$ iff $\mathfrak{A} \models_k \mathbf{A}[v]$ by induction on the structure of $\mathbf{A}$. Most of the cases are simple; we will consider the one in which $\mathbf{A}$ is $\diamond \mathbf{B}$. Now $k \in \mathbf{K} - \mathbf{N}$ iff $k \in h(\mathbf{K}) - (h(\mathbf{K}) \cap \mathbf{N})$, so we may assume that $k \in \mathbf{N}$. Clearly $h(\mathfrak{A}) \models_k \diamond \mathbf{B}[v]$ implies $\mathfrak{A} \models_k \diamond \mathbf{B}[v]$. Now suppose that $\mathfrak{A} \models_k \diamond \mathbf{B}[v]$, so that for some $k' \in \mathbf{K}$ with $k \mathbf{R} k'$, $\mathfrak{A} \models_{k'} \mathbf{B}[v]$. Now for some n, $\mathbf{OR}^n k$, so $\mathbf{OR}^{n+1} k'$, and so $k' \in h(\mathrm{K})$. Then by induction $h(\mathfrak{A}) \models_{k'} \mathbf{B}[v]$ and $h(\mathfrak{A}) \models_k \diamond \mathbf{B}[v]$. ■

Important Convention: In the remainder of this work, unless explicitly stated to the contrary, we will assume that all the structures considered are weakly connected. The difficulty with structures which are not weakly connected appears to be this: A sentence $\mathbf{A}$ of ML is understood as making a statement about a structure $\mathfrak{A}$ *from the point of view of the distinguished world* $\mathbf{O}$: $\mathfrak{A} \models \mathbf{A}$ means $\mathfrak{A} \models_{\mathbf{O}} \mathbf{A}$. This is especially evident in our definitions of validity and elementary equivalence. If $k \in \mathbf{K} - h(\mathbf{K})$, there appears to be no way to use a sentence $\mathbf{A}$ of ML which could make any reference to k or impose any condition upon $\mathscr{A}_k$. Theorems 6.11 and 7.1 show that for weakly connected structures we can use the formulas of ML to impose substantial conditions on the $\mathscr{A}_k$.

We also agree to write $\mathfrak{A} \models_k \mathbf{A}[v]$ only when v is local for $\mathbf{A}$ at k. Note that if v is local for $\mathbf{A}$ at k and $k \mathbf{R} k'$, then v is local for $\mathbf{A}$ at k' since $|\mathscr{A}_k| \subseteq |\mathscr{A}_{k'}|$.

Finally, we say that $\mathfrak{A}$ is a *weak tree* if for all $k \in \mathbf{K}$, the set $\{k' \in \mathbf{K} : k' \mathbf{R} k\}$

is linearly ordered by $\mathbf{R}$, and we say that a system S has the *weak tree property* if every S-theory T which has an S-model has an S-model which is a weak tree. Note that as we use the terminology, a linear ordering is not necessarily anti-symmetric, so that weak trees may contain points k' and k'' with $k' \neq k''$, $k'\mathbf{R}k''$, and $k''\mathbf{R}k'$.

Given $\mathfrak{A} = \langle \mathscr{A}_k, \mathbf{K}, \mathbf{R}, \mathbf{O}, \mathbf{N} \rangle$, define $E(\mathfrak{A}) = \langle \mathscr{B}_l, \mathbf{L}, \mathbf{S}, \mathbf{P}, \mathbf{M} \rangle$ as follows. First set

$$\mathbf{L} = \bigcup_{k \in \mathbf{K}} [b(\mathrm{trc}_k(\mathfrak{A})) \times \mathrm{NAT} \times \{k\}],$$

where NAT is the set of natural numbers. Then define

$$\langle x,n,k \rangle \mathbf{S} \langle y,m,k' \rangle \quad \text{iff} \quad k = k' \;\&\; n = m \;\&\; x\mathbf{R}y,$$

$$\mathbf{P} = \langle \mathbf{O}, 0, \mathbf{O} \rangle,$$

and

$$\mathbf{M} = \{\langle x, n, k \rangle : x \in \mathbf{N}\}.$$

For $\langle x, n, k \rangle = l \in \mathbf{L}$, set $\mathscr{B}_l = \mathscr{A}_x$.

LEMMA 2.19. For any $k, k' \in K$ with $k\mathbf{R}^\infty k'$, the following holds for all formulas $\mathbf{A}$ and assignments v:

(2.20) $\quad \mathfrak{A} \models_{k'} \mathbf{A}[v] \quad$ iff $\quad E(\mathfrak{A}) \models_l \mathbf{A}[v]$,

where $l = \langle k', n, k \rangle$ and $n \in \mathrm{NAT}$.

Proof. This is verified by induction on the formula $\mathbf{A}$, and as in the previous lemma, is simple when $\mathbf{A}$ is atomic or has its principal symbol among $\neg$, $\vee$, or $\exists$. Suppose that $\mathbf{A}$ is $\diamond\mathbf{B}$. If $k' \in \mathbf{K} - \mathbf{N}$, then $l \in \mathbf{L} - \mathbf{M}$, and so by (2.7vi) both sides of (2.20) will hold. Now suppose $k' \in \mathbf{N}$ and hence $l \in \mathbf{M}$. Then by (2.7vi), if $\mathfrak{A} \models_{k'} \diamond\mathbf{B}[v]$, let k'' be such that $k'\mathbf{R}k''$ and $\mathfrak{A} \models_{k''} \mathbf{B}[v]$. Then by the induction hypothesis,

$$E(\mathfrak{A}) \models_{l'} \mathbf{B}[v],$$

where $l' = \langle k'', n, k \rangle$ since $k\mathbf{R}^\infty k''$, and hence

(2.21) $\quad E(\mathfrak{A}) \models_l \diamond\mathbf{B}[v]$.

Now suppose that (2.21) holds. Then by (2.7vi), let $l' \in \mathbf{L}$ be such that $l\mathbf{S}l'$ and

$$E(\mathfrak{A}) \models_{l'} \mathbf{B}[v].$$

Then by the definition of $\mathbf{S}$, there must exist a $k'' \in \mathbf{K}$ such that $l' = \langle k'',n,k \rangle$ and $k'\mathbf{R}k''$. Then by induction, $\mathfrak{A} \models_{k''} \mathbf{B}[v]$ and so $\mathfrak{A} \models_{k'} \mathbf{A}[v]$. ∎

COROLLARY 2.22. The map $\phi : \mathfrak{A} \to E(\mathfrak{A})$ is a strong elementary embedding where $\phi(k) = \langle k, 0, k \rangle$ and $\phi \upharpoonright |\mathscr{A}_k| = \mathrm{id}$.

§3. VALIDITY

We will show here that if S is one of the specific Lemmon or Feys systems considered in §2 then S is valid with respect to Γ_S.

LEMMA 3.1. If S is one of the specific Lemmon or Feys systems considered in §2 and **A** is an axiom of S, then **A** is valid in every S-structure.

Proof. We consider each of the specifically modal axioms listed in §1; we will omit consideration of the axioms and rules for the underlying classical logic. In each case $\mathfrak{A} = \langle \mathscr{A}_k, \mathbf{K}, \mathbf{R}, \mathbf{O}, \mathbf{N} \rangle$ will be an arbitrary S-structure and v will be an arbitrary assignment in $\mathfrak{A}$ and $k \in \mathbf{K}$. We note that it follows from the definition of $\Box$ that

$$\mathfrak{A} \models_k \Box\mathbf{A}[v] \quad \text{iff} \quad k \in \mathbf{K} \text{ and for all } k' \in \mathbf{K} \text{ with } k\mathbf{R}k', \\ \mathfrak{A} \models_{k'} \mathbf{A}[v].$$

First we consider the Lemmon systems.

A0: If $\mathbf{O} \in \mathbf{Q}$, then not $-\mathfrak{A} \models_k \Box(\mathbf{A} \to \mathbf{B})[v]$, so $\mathfrak{A} \models \text{A0}[v]$. If $\mathbf{O} \in \mathbf{N}$, assume $\mathfrak{A} \models \Box(\mathbf{A} \to \mathbf{B}) \& \Box\mathbf{A}[v]$ and assume $\mathbf{OR}k$. Then $\mathfrak{A} \models_k \mathbf{A} \to \mathbf{B}[v]$ and $\mathfrak{A} \models_k \mathbf{A}[v]$, so $\mathfrak{A} \models_k \mathbf{B}[v]$ and hence $\mathfrak{A} \models \Box\mathbf{B}[v]$.

A1: If $\mathbf{O} \in \mathbf{Q}$, not $-\mathfrak{A} \models \Box\mathbf{A}[v]$. If $\mathbf{O} \in \mathbf{N}$, since $\mathfrak{A}$ meets CD1, then $\mathbf{ORO}$, so $\mathfrak{A} \models \mathbf{A}[v]$.

A2: If $\mathbf{O} \in \mathbf{Q}$, not $-\mathfrak{A} \models \Box\mathbf{A}[v]$. If $\mathbf{O} \in \mathbf{N}$, by CD2 there is a $k' \in \mathbf{K}$ with $\mathbf{OR}k'$. Then $\mathfrak{A} \models_{k'} \mathbf{A}[v]$ and so $\mathfrak{A} \models \Diamond\mathbf{A}[v]$.

For most of the remaining axioms we will omit consideration of the case when $\mathbf{O} \in \mathbf{Q}$ since the reasoning is trivial in that case.

A3: Suppose $\mathfrak{A} \models \mathbf{A} \& \Box\mathbf{B}[v]$ and let $\mathbf{OR}k$ (if there is no such k, $\mathfrak{A} \models \Box\Diamond\mathbf{A}[v]$ holds vacuously). If $k \in \mathbf{Q}$, then $\mathfrak{A} \models_k \Diamond\mathbf{A}[v]$. If $k \in \mathbf{N}$, by CD3, $k\mathbf{RO}$, so that $\mathfrak{A} \models_k \Diamond\mathbf{A}[v]$ since $\mathfrak{A} \models \mathbf{A}[v]$. $\mathfrak{A} \models \Box\Diamond\mathbf{A}[v]$ follows.

To condense the presentations we will omit the statement that the antecedents of the axioms hold at **O**.

A4: Let $\mathbf{OR}k$. If $k \in \mathbf{Q}$, not $-\mathfrak{A} \models_k \Box\mathbf{A}[v]$. If $k \in \mathbf{N}$, let $k' \in \mathbf{K}$ with $k\mathbf{R}k'$ and assume $\mathfrak{A} \models_k \Box\mathbf{A}[v]$, so $\mathfrak{A} \models_{k'} \mathbf{A}[v]$. By CD4, $\mathbf{OR}k'$, so $\mathfrak{A} \models_{k'} \mathbf{A} \to \mathbf{B}[v]$ and the result follows.

A5: If $\mathbf{OR}k$, then by CD5, $k \in \mathbf{N}$ and the result follows.

A6: Similar to A5.

A7: Suppose $\mathbf{OR}k$ and $\mathfrak{A}\models_k\mathbf{A}[v]$. Let $k'\in\mathbf{K}$ be arbitrary with $\mathbf{OR}k'$. By CD7, $k'\mathbf{R}k$, so $\mathfrak{A}\models_{k'}\Diamond\mathbf{A}[v]$ and so $\mathfrak{A}\models\Box\Diamond\mathbf{A}[v]$.

A8: Suppose $\mathbf{OR}k$ and assume $\mathfrak{A}\models_k\Box\mathbf{A}[v]$. Since $\mathbf{O}\in\mathbf{N}$ and $\mathbf{OR}k$, by CD8, $k\mathbf{R}k$, and so $\mathfrak{A}\models_k\mathbf{A}[v]$ and hence $\mathfrak{A}\models\Box(\Box\mathbf{A}\rightarrow\mathbf{A})[v]$.

A9: Assume that $\mathfrak{A}\models\mathbf{A}[v]$ and $\mathbf{OR}k$. By CD9, $k=\mathbf{O}$, and so $\mathfrak{A}\models\Box\mathbf{A}[v]$.

A10: If $\mathbf{O}\in\mathbf{Q}$, not $-\mathfrak{A}\models\Box\Box\mathbf{A}[v]$. If $\mathbf{O}\in\mathbf{N}$, by CD10 there is a $k\in\mathbf{Q}$ with $\mathbf{OR}k$. Since not $-\mathfrak{A}\models_k\Box\mathbf{A}[v]$, it follows that $\mathfrak{A}\models\neg\Box\Box\mathbf{A}[v]$.

A11: If $\mathbf{O}\in\mathbf{Q}$, $\mathfrak{A}\models\Diamond\Box\mathbf{A}[v]$. If $\mathbf{O}\in\mathbf{N}$, by CD11 there is a $k\in\mathbf{N}$ with $\mathbf{OR}k$ with no k' such that $k\mathbf{R}k'$. Thus $\mathfrak{A}\models_k\Box\mathbf{A}[v]$ vacuously and so $\mathfrak{A}\models\Diamond\Box\mathbf{A}[v]$.

A12: Since $\mathbf{O}\in\mathrm{Q}$ by CD12, $\mathfrak{A}\models\Diamond\mathbf{A}[v]$.

A13: The second conjunct is treated as for A1. For the first conjunct we may assume $\mathbf{O}\in\mathrm{N}$, and suppose $\mathfrak{A}\models\Box\Diamond\Box\mathbf{A}[v]$. Let $\mathbf{OR}k$. Then $\mathfrak{A}\models_k\Diamond\Box\mathbf{A}[v]$, so there is a $k'\in\mathbf{K}$ with $k\mathbf{R}k'$ and $\mathfrak{A}\models_{k'}\Box\mathbf{A}[v]$. But by CD13, $k'\mathbf{R}k$ and so we must have $\mathfrak{A}\models_k\mathbf{A}[v]$, and so $\mathfrak{A}\models\Box\mathbf{A}[v]$.

A14: Suppose $\mathfrak{A}\models\Diamond^m\Box^n\mathbf{A}[v]$ so that since $\mathbf{O}\in\mathbf{N}$, for some $k\in\mathbf{K}$ with $\mathbf{OR}^m k$, $\mathfrak{A}\models_k\Box^n\mathbf{A}[v]$. Let $k'\in\mathbf{K}$ with $\mathbf{OR}^p k'$. By CD14, there is a $k''\in\mathbf{K}$ with $k\mathbf{R}^n k''$ and $k'\mathbf{R}^q k''$. Since $\mathfrak{A}\models_k\Box^n\mathbf{A}[v]$ then $\mathfrak{A}\models_{k''}\mathbf{A}[v]$ and since $k'\mathbf{R}^q k''$, it follows that $\mathfrak{A}\models_{k'}\Diamond^q\mathbf{A}[v]$, so $\mathfrak{A}\models\Box^p\Diamond^q\mathbf{A}[v]$.

Next we consider the Feys systems. For axioms A1′–A7′, A10′ the reasoning is quite similar to that for A1–A7, A10 respectively, and so we omit the details. Recall that in this context we always have $\mathbf{O}\in\mathbf{N}$.

A8′: By CD8′ there is a $k\in\mathbf{Q}$ with $\mathbf{OR}k$. Then $\mathfrak{A}\models_k\Diamond\mathbf{A}[v]$ and so $\mathfrak{A}\models\Diamond\Diamond\mathbf{A}[v]$.

A9′: By CD9′ there is a $k\in\mathbf{N}$ with $\mathbf{OR}k$ such that for no $k'\in\mathbf{K}$ do we have $k\mathbf{R}k'$. Then $\mathfrak{A}\models_k\Box\mathbf{A}[v]$ holds vacuously and so $\mathfrak{A}\models\Diamond\Box\mathbf{A}[v]$.

For □A1′–□A10′ the reasoning is similar to that for A1′–A10′. We consider one example.

□A7′: Recall that $\mathbf{O}\in\mathbf{N}$ and let $k\in\mathbf{K}$ be such that $\mathbf{OR}k$. If $k\in\mathbf{Q}$, not $-\mathfrak{A}\models_k\Box\mathbf{B}[v]$, so assume $k\in\mathbf{N}$ and that $\mathfrak{A}\models_k\Box\mathbf{B}[v]$ and $\mathfrak{A}\models_k\Diamond\mathbf{A}[v]$. Then for some $k'\in\mathbf{K}$ with $k\mathbf{R}k'$, we have $\mathfrak{A}\models_{k'}\mathbf{A}[v]$. Let $k''\in\mathbf{K}$ be such that $k\mathbf{R}k''$. If $k''\in\mathbf{Q}$, then $\mathfrak{A}\models_{k''}\Diamond\mathbf{A}[v]$. If $k''\in\mathbf{N}$, by CD7, since $k\in\mathbf{N}$, $k\mathbf{R}k'$, and $k\mathbf{R}k''$, then $k''\mathbf{R}k'$. But this yields $\mathfrak{A}\models_{k''}\Diamond\mathbf{A}[v]$, so $\mathfrak{A}\models_k\Box\Diamond\mathbf{A}[v]$, as desired. This completes the proof of the Lemma.

LEMMA 3.2. If S is one of the specific Lemmon or Feys systems considered in §2, if $\mathfrak{A}=\langle\mathscr{A},\mathbf{K},\mathbf{R},\mathbf{O},\mathbf{N}\rangle$ is an S-structure and $k\in\mathbf{K}$, then $\mathrm{trc}_k(\mathfrak{A})$ is again an S-structure, provided that when S is a Feys system, $k\in\mathbf{N}$.

Proof. It is clear that $\mathrm{trc}_k(\mathfrak{A})$ is again a modal structure. Now if **C** is one of CD1, CD3–CD9, CD12, CD13, or Norm, then **C** is equivalent to a universal sentence. Now $b(\mathfrak{A}) \models \mathbf{C}$ and $b(\mathrm{trc}_k(\mathfrak{A}))$ is a substructure of $b(\mathfrak{A})$. Hence by the Łos-Tarski Theorem (Shoenfield, 1967, p. 76), $b(\mathrm{trc}_k(\mathfrak{A})) \models \mathbf{C}$. If **C** is CD2, CD10, CD11, or CD14, **C** is equivalent to a sentence of the form $\forall \mathbf{x}_1 \ldots \forall \mathbf{x}_n \exists \mathbf{y}[\mathbf{x}_i \mathbf{R}^* \mathbf{y} \,\&\, \mathbf{D}]$, where **D** is open. It then follows from the definition of truncation and the Łos-Tarski Theorem that $b(\mathrm{trc}_k(\mathfrak{A})) \models \mathbf{C}$. ■

LEMMA 3.3 If **A** is a formula of ML which is a classical theorem (cf. §1), then **A** is valid in any modal structure.

Proof. We proceed by induction on the length of the given proof of **A**. Note that the only axioms used in this proof are instances of axioms of the classical predicate calculus and the only rules used are non-modal rules. The details are easy but tedious, so we omit them. ■

CONSISTENCY LEMMA (3.4). If S is one of the specific Lemmon or Feys systems considered in §2 and if $\vdash^{S} \mathbf{A}$, then **A** is valid in every S-structure.

Proof. We proceed by induction on the length of the given proof of **A** in S. By virtue of Lemma 3.1 it suffices to show that the property of being valid in every S-structure is preserved by the rules of inference of S. This is simple for the non-modal rules. Let $\mathfrak{A} = \langle \mathscr{A}_k, \mathbf{K}, \mathbf{R}, \mathbf{O}, \mathbf{N} \rangle$ be a fixed S-structure and let v be an arbitrary assignment in $\mathfrak{A}$. We first consider the Lemmon systems.

R2: Suppose that $\mathbf{A} \rightarrow \mathbf{B}$ is valid in all S-structures. If $\mathbf{O} \in \mathbf{Q}$, then not $-\mathfrak{A} \models \Box \mathbf{A}[v]$, so assume $\mathbf{O} \in \mathbf{N}$ and that $\mathfrak{A} \models \Box \mathbf{A}[v]$, and let $k \in \mathbf{K}$ be such that $\mathbf{OR}k$. Then $\mathfrak{A} \models_k \mathbf{A}[v]$. Now it is easy to see that for any **A**,

$$\mathfrak{A} \models_k \mathbf{A}[v] \quad \text{iff} \quad \mathrm{trc}_k(\mathfrak{A}) \models \mathbf{A}[v].$$

Now by Lemma 3.2, $\mathrm{trc}_k(\mathfrak{A})$ is again an S-structure and so by hypothesis, $\mathfrak{A} \models_k \mathbf{A} \rightarrow \mathbf{B}[v]$. Hence $\mathfrak{A} \models_k \mathbf{B}[v]$, as desired.

R3: Suppose that **A** is valid in all S-structures and let $k \in \mathbf{K}$. By Lemma 3.2, $\mathrm{trc}_k(\mathfrak{A})$ is again an S-structure and so $\mathfrak{A} \models_k \mathbf{A}[v]$. Since S includes R3, then $\mathbf{N} = \mathbf{K}$, so $\mathbf{O} \in \mathbf{N}$, and hence $\mathfrak{A} \models \Box \mathbf{A}[v]$.

Now we consider the Feys systems.

R2′: Suppose $\Box(\mathbf{A} \rightarrow \mathbf{B})$ is valid in all S-structures. Since $\mathfrak{A}$ is now a Feys structure, $\mathbf{O} \in \mathbf{N}$. Let $k \in \mathbf{K}$ with $\mathbf{OR}k$. If $k \in \mathbf{Q}$, then not $-\mathfrak{A} \models_k \Box \mathbf{A}[v]$, so assume that $\mathfrak{A} \models_k \Box \mathbf{A}[v]$ with $k \in \mathbf{N}$. Since $k \in \mathbf{N}$, then $\mathrm{trc}_k(\mathfrak{A})$ is an S-structure and so $\mathfrak{A} \models_k \Box(\mathbf{A} \rightarrow \mathbf{B})[v]$ by hypothesis. Thus for any k' with

$k\mathbf{R}k'$ we have $\mathfrak{A}\models_{k'}\mathbf{A}\,\&\,\mathbf{A}\rightarrow\mathbf{B}[v]$. It follows that $\mathfrak{A}\models_{k}\Box\mathbf{B}[v]$ and so $\mathfrak{A}\models\Box(\Box\mathbf{A}\rightarrow\Box\mathbf{B})[v]$.

$\mathrm{R3}_t$: Suppose **A** is classical theorem and $k\in\mathbf{K}$. By Lemma 3.3, $\mathfrak{A}\models_{k}\mathbf{A}[v]$ and so $\mathfrak{A}\models\Box\mathbf{A}[v]$. This completes the proof. ■

VALIDITY THEOREM (3.5). Let S be one of the Lemmon or Feys systems specifically considered and let T be an S-theory. If $\vdash^{S}_{T}\mathbf{A}$, then **A** is valid in every S-model of T.

Proof. Let $\mathfrak{A}$ be an S-model of T; in particular, $\mathfrak{A}$ is an S-structure. Since $\vdash^{S}_{T}\mathbf{A}$, there are nonlogical axioms $\mathbf{B}_1,\ldots,\mathbf{B}_n$ of T such that $\vdash^{S}\neg\mathbf{B}_1\vee\ldots\vee\neg\mathbf{B}_n\vee\mathbf{A}$. By the Consistency Lemma, $\neg\mathbf{B}_1\vee\ldots\vee\neg\mathbf{B}_n\vee\mathbf{A}$ is valid in $\mathfrak{A}$. But since $\mathfrak{A}$ is an S-model of T, $\mathbf{B}_i$ is valid in $\mathfrak{A}$ for $i=1,\ldots,n$, and so **A** must be valid in $\mathfrak{A}$. ■

§4. COMPLETENESS

Our approach to the completeness problem is in the style of Henkin as developed by Lemmon and Scott (1966), Makinson (1966), Routley (1970), Schutte (1970), van Fraasen (1969), and Thomason (1970) (cf. also Aczel (1968), Åqvist (1971), Kripke (1959), and Thomason (1968)). An S-theory T is said to be S-*consistent* if there are no nonlogical axioms $\mathbf{B}_1, \ldots, \mathbf{B}_n$ of T such that

$$(4.1) \qquad \vdash^{\mathrm{S}}_{\mathrm{T}} \neg \mathbf{B}_1 \vee \ldots \vee \neg \mathbf{B}_n .$$

This is clearly equivalent to the assertion that there is no formula $\mathbf{A}$ such that $\vdash^{\mathrm{S}}_{\mathrm{T}} \mathbf{A} \wedge \neg \mathbf{A}$. An S-theory T in the language ML is said to be S-*complete* iff T is S-consistent and for every sentence $\mathbf{A}$ of ML, either $\vdash^{\mathrm{S}}_{\mathrm{T}} \mathbf{A}$ or $\vdash^{\mathrm{S}}_{\mathrm{T}} \neg \mathbf{A}$. An S-theory T in the language ML is called a *Henkin* theory iff for every formula $\mathbf{A}$ of ML whose only free variable is $\mathbf{x}$, there is an individual constant $\mathbf{c}$ of ML such that

$$(4.2) \qquad \vdash^{\mathrm{S}}_{\mathrm{T}} \exists \mathbf{x} \mathbf{A} \rightarrow \mathbf{A}_{\mathbf{x}}[\mathbf{c}].$$

Finally, the S-theory T′ is said to be an *extension* of the S-theory T if the language of T′ includes the language of T and for every formula $\mathbf{A}$ of the language of T, $\vdash^{\mathrm{S}}_{\mathrm{T}} \mathbf{A}$ implies $\vdash^{\mathrm{S}}_{\mathrm{T}'} \mathbf{A}$.

LINDENBAUM'S LEMMA (4.3). Any S-consistent S-theory T can be extended to an S-complete theory.

Proof. Just as in the classical case, let $\mathscr{T}$ be the set of all S-consistent extensions of T in the same language as T. Then $\mathrm{T} \in \mathscr{T}$ and it is easy to see that the union of any chain in $\mathscr{T}$ under $\subseteq$ is again an S-consistent extension of T and so by Zorn's Lemma (cf. Halmos (1960) or Kuratowski and Mostowski (1968)), $\mathscr{T}$ contains a maximal element, call one such T′. Then T′ is S-consistent. If $\mathrm{T}'[\mathbf{A}]$ is S-inconsistent, then $\vdash^{\mathrm{S}}_{\mathrm{T}'} \neg \mathbf{A}$. Hence if neither $\vdash^{\mathrm{S}}_{\mathrm{T}'} \mathbf{A}$ nor $\vdash^{\mathrm{S}}_{\mathrm{T}'} \neg \mathbf{A}$, then both $\mathrm{T}'[\mathbf{A}]$ and $\mathrm{T}'[\neg \mathbf{A}]$ would be S-consistent, contradicting the maximality of T′. Thus T′ is S-complete. ■

It is easy to see in fact that T is a maximal S-consistent S-theory iff T is S-complete.

LEMMA 4.4. Let S be a Lemmon system and let T be an S-consistent S-theory such that for some $\mathbf{B}$, $\vdash^{S}_{T}\Box\mathbf{B}$. Suppose $\vdash^{S}_{T}\Diamond\mathbf{A}$ and let $d_{\mathbf{A}}T$ be the S-theory whose nonlogical axioms consist of $\mathbf{A}$ and all $\mathbf{C}$ such that $\vdash^{S}_{T}\Box\mathbf{C}$. Then $d_{\mathbf{A}}T$ is S-consistent.

Proof. Suppose $d_{\mathbf{A}}T$ were not S-consistent. Then there would be formulas $\mathbf{C}_1, \ldots, \mathbf{C}_n$ such that $\vdash^{S}_{T}\Box\mathbf{C}_i$ for $i = 1, \ldots, n$ and

$$\vdash^{S}\neg\mathbf{C}_1 \vee \ldots \vee \neg\mathbf{C}_n \vee \neg\mathbf{A}.$$

If $n = 0$ we have $\vdash^{S}\neg\mathbf{A}$ so $\vdash^{S}\mathbf{B}\rightarrow\neg\mathbf{A}$ and so by R2 $\vdash^{S}\Box\mathbf{B}\rightarrow\Box\neg\mathbf{A}$. Then by tautology it follows that $\vdash^{S}_{T}\Box\neg\mathbf{A}$ and then by definition of $\Box$, $\vdash^{S}_{T}\neg\Diamond\neg\neg\mathbf{A}$. Then by Theorem 1.10, $\vdash^{S}_{T}\neg\mathbf{A}$, contradicting the S-consistency of T.

Suppose $n > 0$. Then by R2 and repeated use of axiom A0 and tautology, we have

$$\vdash^{S}_{T}\Box\mathbf{C}_1 \rightarrow\centerdot\ \Box\mathbf{C}_2 \rightarrow\centerdot\ \ldots \rightarrow\centerdot\ \Box\mathbf{C}_n \rightarrow \Box\neg\mathbf{A}.$$

Since $\vdash^{S}_{T}\Box\mathbf{C}_i$ for $i = 1, \ldots, n$, then by tautology, $\vdash^{S}_{T}\Box\neg\mathbf{A}$, and as above, we have $\vdash^{S}_{T}\neg\Diamond\mathbf{A}$, again contradicting the S-consistency of T. ■

For any S, U, U′, and $n \geq 0$, we let $d^n(\mathrm{U}, \mathrm{U}')$ be the S-theory whose set of non-logical axioms is

$$\{\mathbf{A}: \vdash^{S}_{U}\Box\mathbf{A}\} \cup \{\Diamond^n\mathbf{B}: \vdash^{S}_{U'}\mathbf{B}\},$$

and we let $e^{n,q}$ be the S-theory whose set of non-logical axioms is

$$\{\mathbf{A}: \vdash^{S}_{U}\Box^n\mathbf{A}\} \cup \{\mathbf{A}: \vdash^{S}_{U'}\Box^q\mathbf{A}\}.$$

For the following, let CS be the system with no non-modal axioms or rules, i.e., the theoremsof CS are what we called classical theorems in §1.

LEMMA 4.5. Let S be a modal system containing A0 and closed under $\mathrm{R3}_\iota$, and let T be an S-consistent S-theory such that $\vdash^{S}_{T}\Diamond\mathbf{A}$. Let $d_{\mathbf{A}}T$ be the S-theory whose nonlogical axioms are $\mathbf{A}$ together with all $\mathbf{B}$ such that $\vdash^{S}_{T}\Box\mathbf{B}$. Then $d_{\mathbf{A}}T$ is CS-consistent.

Proof. If $d_{\mathbf{A}}T$ is not CS-consistent there are $\mathbf{B}_1, \ldots, \mathbf{B}_n$ with $n \geq 0$ such that for $i = 1, \ldots, n$, $\vdash^{S}_{T}\mathbf{B}_i$, and

$$\vdash^{CS}\neg\mathbf{B}_1 \vee \ldots \vee \neg\mathbf{B}_n \vee \neg\mathbf{A}. \tag{4.6}$$

If $n = 0$, then $\vdash^{CS}\neg\mathbf{A}$, so $\vdash^{S}\Box\neg\mathbf{A}$ and hence $\vdash^{S}\neg\Diamond\neg\neg\mathbf{A}$. Since $\vdash^{CS}\mathbf{A}\leftrightarrow\neg\neg\mathbf{A}$, then $\vdash^{S}\Box(\mathbf{A}\leftrightarrow\neg\neg\mathbf{A})$, and so by Theorem 1.8, $\vdash^{S}\neg\Diamond\mathbf{A}$ and so $\vdash^{S}_{T}\neg\Diamond\mathbf{A}$, contradicting the S-consistency of T.

If $n > 0$, by (4.6) and $R3_t$ it follows that

$$\vdash^S_T \Box(\mathbf{B}_1 \to \centerdot\ \mathbf{B}_2 \to \centerdot \ldots \to \centerdot\ \mathbf{B}_n \to \neg \mathbf{A}).$$

Then by repeated use of A0 and tautology,

$$\vdash^S_T \Box\mathbf{B}_1 \to \centerdot\ \Box\mathbf{B}_2 \to \centerdot \ldots \to \centerdot\ \Box\mathbf{B}_n \to \Box\neg \mathbf{A}.$$

Since $\vdash^S_T \Box\mathbf{B}_i$ for $i = 1, \ldots, n$, then $\vdash^S_T \Box\neg\mathbf{A}$, and as above, $\vdash^S_T \neg\Diamond\mathbf{A}$, contradicting the S-consistency of T. ■

LEMMA 4.7. Let S be a modal system and let T be an S-consistent S-theory. Then T can be consistently extended in S to a Henkin theory.

Proof. We adapt the method of Shoenfield. Let ML be the language of T. We define the *special constants of level n with* S *and* T by induction on n as follows. Assume these have been defined for $n < m$ and let ML_{m-1} be the language obtained by adding all these special constants to ML as new individual constants. Let $\exists x\mathbf{A}$ be any closed formula of ML_{m-1} such that if $m > 0$, then $\mathbf{A}$ contains at least one special constant of level $m - 1$. Then the symbol

(4.8) $\quad C^{S,T}_{\exists x\mathbf{A}}$

is a special constant of level m wrt S and T; it is called the special constant *for* the formula $\exists x\mathbf{A}$. Then formula

(4.9) $\quad \exists \mathbf{x}\mathbf{A} \to \mathbf{A}_{\mathbf{x}}[e]$,

where e is the special constant (4.8), is called the *special axiom* for the special constant (4.8).

Let $ML^{S,T}_C$ be the language obtained from ML by adding the special constants of level n wrt S and T as new individual constants for all $n > 0$. Let T_C be the S-theory formulated in ML_C whose axioms are those of T together with all the special axioms (4.9) for all the special constants (4.8) of all levels. Let T′ be the S-theory whose language is ML_C and whose non-logical axioms are just those of T. Let $\mathbf{A}$ be a formula of ML and suppose that $\vdash^S_{T_C} \mathbf{A}$. Then there must be special axioms $\mathbf{B}_1, \ldots, \mathbf{B}_n$ and nonlogical axioms $\mathbf{C}_1, \ldots, \mathbf{C}_m$ of T such that

$$\vdash^S \mathbf{C}_1 \to \centerdot \ldots \to \centerdot\ \mathbf{C}_m \to \centerdot\ \mathbf{B}_1 \to \centerdot\ \mathbf{B}_2 \to \ldots \to \centerdot\ \mathbf{B}_n \to \mathbf{A}.$$

Then we get

$$\vdash^S \mathbf{B}_1 \to \centerdot\ \mathbf{C}_1 \to \centerdot \ldots \to \centerdot\ \mathbf{C}_m \to \centerdot\ \mathbf{B}_2 \to \ldots \to \centerdot\ \mathbf{B}_n \to \mathbf{A}.$$

We may assume that the level of the special constant $\mathbf{C}_1$ for which $\mathbf{B}_1$

is a special axiom is at least as great as the special constants for which $\mathbf{B}_2, \ldots, \mathbf{B}_n$ are special axioms. It follows from the definition of special constant and axiom that:

(a) $\mathbf{B}_1$ is of the form $\exists \mathbf{x}\mathbf{C} \to \mathbf{C}_{\mathbf{x}}[c_1]$,

(b) c_1 does not occur in $\mathbf{C}, \mathbf{B}_2, \ldots, \mathbf{B}_n, \mathbf{A}$, or $\mathbf{C}_1, \ldots, \mathbf{C}_m$. Then by the Theorem on Constants (1.6), if $\mathbf{y}$ is a new variable not occurring in $\mathbf{B}_1, \ldots, \mathbf{B}_n, \mathbf{A}$, or $\mathbf{C}_1, \ldots, \mathbf{C}_m$,

$$\vdash^{\mathrm{S}} (\exists \mathbf{x}\mathbf{C} \to \mathbf{C}_{\mathbf{x}}[\mathbf{y}]) \to \blacksquare\, \mathbf{C}_1 \to \blacksquare \ldots \to \blacksquare\, \mathbf{C}_m \to \blacksquare\, \mathbf{B}_2 \to \blacksquare \ldots \to$$
$$\to \blacksquare\, \mathbf{B}_n \to \mathbf{A}.$$

Then by the $\exists$-Introduction Rule,

$$\vdash^{\mathrm{S}} \exists \mathbf{y}(\exists \mathbf{x}\mathbf{C} \to \mathbf{C}_{\mathbf{x}}[\mathbf{y}]) \to \blacksquare\, \mathbf{C}_1 \to \blacksquare \ldots \to \blacksquare\, \mathbf{C}_m \to \blacksquare\, \mathbf{B}_2 \to \blacksquare \ldots$$
$$\to \blacksquare\, \mathbf{B}_n \to \mathbf{A}.$$

But then by the classical prenex operations and change of bound variable, $\vdash^{\mathrm{S}} \exists \mathbf{y}(\exists \mathbf{x}\mathbf{C} \to \mathbf{C}_{\mathbf{x}}[\mathbf{y}])$, so that

$$\vdash^{\mathrm{S}}_{\mathrm{T}'} \mathbf{B}_2 \to \blacksquare \ldots \to \blacksquare\, \mathbf{B}_n \to \mathbf{A}.$$

By induction on n, it follows that $\vdash^{\mathrm{S}}_{\mathrm{T}'} \mathbf{A}$ and then once more by the Theorem on Constants (1.6), $\vdash^{\mathrm{S}}_{\mathrm{T}} \mathbf{A}$. Thus T_{C} is a *conservative extension* of T in the sense that for any formula $\mathbf{A}$ of ML, $\vdash^{\mathrm{S}}_{\mathrm{T}_{\mathrm{C}}} \mathbf{A}$ iff $\vdash^{\mathrm{S}}_{\mathrm{T}} \mathbf{A}$. Then if T_{C} were inconsistent, for any formula $\mathbf{A}$ of ML we would have $\vdash^{\mathrm{S}}_{\mathrm{T}_{\mathrm{C}}} \mathbf{A} \,\&\, \neg \mathbf{A}$ and so $\vdash^{\mathrm{S}}_{\mathrm{T}} \mathbf{A} \,\&\, \neg \mathbf{A}$, contradicting the S-consistency of T. It is clear that T_{C} is a Henkin theory. ■

In the following, let $^{+}$ be a function on S-theories such that for any S-consistent S-theory T, T^{+} is an S-complete extension of T. We write $\mathrm{T}^{+}_{\mathrm{C}}$ for $(\mathrm{T}_{\mathrm{C}})^{+}$.

DEFINITION 4.10. Let S be a Lemmon system and let T be S-consistent. We define the *canonical structure for* T, $\mathfrak{A} = \langle \mathscr{A}_k, \mathbf{K}, \mathbf{R}, \mathbf{O}, \mathbf{N} \rangle$, as follows. Let $\mathbf{K}$ be the smallest set of S-theories such that $\mathrm{T}^{+}_{\mathrm{C}} \in \mathbf{K}$ and

(i) $\mathrm{U} \in \mathbf{K} \,\&\, \vdash^{\mathrm{S}}_{\mathrm{U}} \Diamond \mathbf{A} \,\&$ for some $\mathbf{B}$, $\vdash^{\mathrm{S}}_{\mathrm{U}} \Box \mathbf{B} \Rightarrow (d_{\mathbf{A}}\mathrm{U})^{+}_{\mathrm{C}} \in \mathbf{K}$;

(ii) $\mathrm{U}, \mathrm{U}' \in \mathbf{K} \,\&\, n \geq 0 \,\&\, \{\mathbf{A} : \vdash^{\mathrm{S}}_{\mathrm{U}} \Box^{n+1} \mathbf{A}\} \subseteq \mathrm{U}' \Rightarrow (d^{n}(\mathrm{U}, \mathrm{U}'))^{+}_{\mathrm{C}} \in \mathbf{K}$;

(iii) $n, q \geq 0 \,\&\, \mathrm{U}, \mathrm{U}' \in \mathbf{K} \Rightarrow (e^{n,q}(\mathrm{U}, \mathrm{U}'))^{+}_{\mathrm{C}} \in \mathbf{K}$.

Note that $\mathrm{card}(\mathbf{K}) \leq \mathrm{card}(\mathrm{ML}(\mathrm{T}))$ where the latter is the cardinality of the set of all formulas of ML(T). We set $\mathbf{O} = \mathrm{T}^{+}_{\mathrm{C}}$, set

$$\mathbf{N} = \{\mathrm{U} \in \mathbf{K} : \text{for some } \mathbf{B}, \vdash^{\mathrm{S}}_{\mathrm{U}} \Box \mathbf{B}\},$$

and for U, $U' \in \mathbf{K}$, we define

$$\mathrm{U}\mathbf{R}\mathrm{U}' \text{ iff } \{\mathbf{B}\colon \vdash^{\mathrm{S}}_{\mathrm{U}} \Box\mathbf{B}\} \subseteq \mathrm{U}' \;\&\; \text{for some } \mathbf{B}, \vdash^{\mathrm{S}}_{\mathrm{U}} \Box\mathbf{B}.$$

Note that we always have $\mathrm{U}\mathbf{R}(d_{\mathrm{A}}\mathrm{U})^{+}_{\mathrm{C}}$.

To define the $\mathscr{A}_k$ for $k = \mathrm{U} \in \mathbf{K}$, we proceed as follows. $|\mathscr{A}_k|$ consists of all closed terms of ML(k), where $k \in \mathbf{K}$. For $\mathbf{e}, \mathbf{e}' \in |\mathscr{A}_k|$, we define $\mathbf{e} =_k \mathbf{e}'$ iff $\vdash^{\mathrm{S}}_k \mathbf{e} = \mathbf{e}'$. Moreover, if $\mathbf{f}$ is an n-ary function symbol of ML and $\mathbf{e}_1, \ldots, \mathbf{e}_n \in |\mathscr{A}_k|$ we set $\mathbf{f}_{\mathscr{A}_k}(\mathbf{e}_1, \ldots, \mathbf{e}_n)$ equal to $\mathbf{f}\mathbf{e}_1 \ldots \mathbf{e}_n$. Finally, if $\mathbf{p}$ is an n-ary predicate symbol of ML, then define $\mathbf{p}_{\mathscr{A}_k}(\mathbf{e}_1, \ldots, \mathbf{e}_n)$ iff $\vdash^{\mathrm{S}}_k \mathbf{p}\mathbf{e}_1 \ldots \mathbf{e}_n$. It is easy to check that $=_k$ is an equivalence relation on $|\mathscr{A}_k|$ and is in fact a congruence for each $\mathbf{f}_{\mathscr{A}_k}$ and $\mathbf{p}_{\mathscr{A}_k}$. This completes the specification of $\mathfrak{A}$. It is easy to see that if $k\mathbf{R}k'$, then $|\mathscr{A}_k| \subseteq |\mathscr{A}_{k'}|$, so that $\mathfrak{A}$ is a modal structure.

Let us first make an observation which will prove important later. Suppose that S is a semi-normal system and let U, U′, and $\mathrm{U}'' \in \mathbf{K}$ where $\mathrm{U}'\mathbf{R}\mathrm{U}$ and $\mathrm{U}''\mathrm{R}\mathrm{U}$. Let $\mathbf{B}$ be a closed formula of the common language of U′ and U″ such that $\vdash_{\mathrm{U}'}\mathbf{B}$. Then by R3, $\vdash_{\mathrm{U}'}\Box\mathbf{B}$, and since $\mathrm{U}'\mathbf{R}\mathrm{U}$, then $\vdash_{\mathrm{U}}\mathbf{B}$. Suppose also that $\vdash_{\mathrm{U}''}\neg\mathbf{B}$. Then also by R3, $\vdash_{\mathrm{U}''}\Box\neg\mathbf{B}$, and since $\mathrm{U}''\mathbf{R}\mathrm{U}$, $\vdash_{\mathrm{U}}\neg\mathbf{B}$, contradicting the S-consistency of U. Thus not $-\vdash_{\mathrm{U}''}\neg\mathbf{B}$, and since U″ is S-complete, $\vdash_{\mathrm{U}''}\mathbf{B}$. Thus $\mathrm{U}'\mathbf{R}\mathrm{U}''$. Similarly, $\mathrm{U}''\mathbf{R}\mathrm{U}'$. Thus the canonical structure is a weak tree when S is semi-normal.

LEMMA 4.11. Let S be one of the Lemmon systems specifically considered in §2, and let T be an S-consistent S-theory. Then the canonical structure for T is an S-structure.

Proof. We consider the cases of the various axioms and rules in turn.

(A1). Suppose A1 is an axiom of S; we must show that $\mathbf{R}$ is reflexive. Let $\mathrm{U} \in \mathbf{K}$. Then for any formula $\mathbf{B}$, $\vdash^{\mathrm{S}}_{\mathrm{U}} \Box\mathbf{B} \to \mathbf{B}$. Thus if $\mathbf{B}$ is an arbitrary formula and $\vdash^{\mathrm{S}}_{\mathrm{U}} \Box\mathbf{B}$, then $\vdash^{\mathrm{S}}_{\mathrm{U}}\mathbf{B}$, and so $\mathrm{U}\mathbf{R}\mathrm{U}$.

Henceforth in each case we will omit the explicit statement that the sentence A_i in question is an axiom of S and that the condition CD_i is to be verified.

(A2) Let $\mathrm{U} \in \mathbf{N}$ so that for some formula $\mathbf{B}$, $\vdash^{\mathrm{S}}_{\mathrm{U}} \Box\mathbf{B}$. Then using A2, $\vdash^{\mathrm{S}}_{\mathrm{U}} \Diamond\mathbf{B}$. Let $\mathrm{U}' = (d_{\mathbf{B}}\mathrm{U})^{+}_{\mathrm{C}}$. Then $\mathrm{U}' \in \mathbf{K}$ and $\mathrm{U}\mathbf{R}\mathrm{U}'$ as desired.

(A3) Let $\mathrm{U}, \mathrm{U}' \in \mathbf{N}$ and suppose that $\mathrm{U}\mathbf{R}\mathrm{U}'$. Let $\mathbf{A}$ be such that $\vdash^{\mathrm{S}}_{\mathrm{U}'} \Box\mathbf{A}$, and suppose not $-\vdash^{\mathrm{S}}_{\mathrm{U}}\mathbf{A}$. Then since U is S-complete, $\vdash^{\mathrm{S}}_{\mathrm{U}}\neg\mathbf{A}$. Now since $\mathrm{U} \in \mathbf{N}$, then for some $\mathbf{B}$, $\vdash^{\mathrm{S}}_{\mathrm{U}} \Box\mathbf{B}$. Then by A3, $\vdash^{\mathrm{S}}_{\mathrm{U}} \neg\mathbf{A} \to \Box\Diamond\neg\mathbf{A}$, and so $\vdash^{\mathrm{S}}_{\mathrm{U}} \Box\Diamond\neg\mathbf{A}$. Since $\mathrm{U}\mathbf{R}\mathrm{U}'$, then $\vdash^{\mathrm{S}}_{\mathrm{U}'} \Diamond\neg\mathbf{A}$ and hence $\vdash^{\mathrm{S}}_{\mathrm{U}'} \neg\Box\mathbf{A}$, contradicting the S-consistency of U′ using Lemma 4.4. Hence we must have $\vdash^{S}_{U}\mathbf{A}$ and hence $U'\mathbf{R}U$.

(A4) Let $U, V \in \mathbf{N}$ and $W \in \mathbf{K}$ and suppose $U\mathbf{R}V$ and $V\mathbf{R}W$. Let $\mathbf{A}$ be such that $\vdash^{S}_{U}\Box\mathbf{A}$. Now for some $\mathbf{B}$, $\vdash^{S}_{V}\Box\mathbf{B}$, and by Lemma 1.13, $\vdash^{S}_{U}\Box\mathbf{A}\rightarrow\Box(\Box\mathbf{B}\rightarrow\Box\mathbf{A})$, so that $\vdash^{S}_{U}\Box(\Box\mathbf{B}\rightarrow\Box\mathbf{A})$. Thus $\vdash^{S}_{V}\Box\mathbf{B}\rightarrow\Box\mathbf{A}$, so that $\vdash^{S}_{V}\Box\mathbf{A}$ and hence $\vdash^{S}_{W}\mathbf{A}$. Thus $U\mathbf{R}U$.

(A5) Let $U \in \mathbf{N}$, $U' \in \mathbf{K}$ and $U\mathbf{R}U'$. Since $U \in \mathbf{N}$, then for some $\mathbf{B}$, $\vdash^{S}_{U}\Box\mathbf{B}$. Now $\vdash^{S}\mathbf{B}\rightarrow x = x$, so by R2, $\vdash^{S}_{U}\Box\mathbf{B}\rightarrow\Box x = x$, and so $\vdash^{S}_{U}\Box x = x$. Then by A5, $\vdash^{S}_{U}\Box\Box x = x$ and hence $\vdash^{S}_{U'}\Box x = x$ and so $U' \in \mathbf{N}$.

(A6) To show $C1_{N}$, note that $\vdash^{S}$ A5, and proceed as above. For $Tran_{N}$, suppose $U, V \in \mathbf{N}$, $W \in \mathbf{K}$, $U\mathbf{R}V$ and $V\mathbf{R}W$. Let $\vdash^{S}_{U}\Box\mathbf{A}$. By A6, $\vdash^{S}_{U}\Box\Box\mathbf{A}$, $\vdash^{S}_{V}\Box\mathbf{A}$, so $\vdash^{S}_{W}\mathbf{A}$, and thus $U\mathbf{R}W$.

(A7) Let $U \in \mathbf{N}$, $V, W \in \mathbf{K}$, $U\mathbf{R}V$, and $U\mathbf{R}W$. Let $\vdash^{S}_{V}\Box\mathbf{A}$. Suppose not $-\vdash^{S}_{W}\mathbf{A}$. Then not $-\vdash^{S}_{U}\Box\mathbf{A}$ and since U is S-complete, this yields $\vdash^{S}_{U}\neg\Box\mathbf{A}$ and so $\vdash^{S}_{U}\Diamond\neg\mathbf{A}$. Now since $U \in \mathbf{N}$, then for some $\mathbf{B}$, $\vdash^{S}_{U}\Box\mathbf{B}$, and using A7, we get $\vdash^{S}_{U}\Diamond\neg\mathbf{A}\rightarrow\Box\Diamond\neg\mathbf{A}$, and hence $\vdash^{S}_{U}\Box\Diamond\neg\mathbf{A}$. Thus $\vdash^{S}_{V}\Diamond\neg\mathbf{A}$ and so $\vdash^{S}_{V}\neg\Box\mathbf{A}$, contradicting the S-consistency of V, using Lemma 4.4.

(A8). Let $U \in \mathbf{K}$ and suppose that $U' \in \mathbf{N}$ and $U'\mathbf{R}U$. Let $\vdash^{S}_{U}\Box\mathbf{A}$. Since $U' \in \mathbf{N}$, then $\vdash^{S}_{U'}\Box\mathbf{B}$ for some $\mathbf{B}$. Using A8, $\vdash^{S}_{U'}\Box(\Box\mathbf{A}\rightarrow\mathbf{A})$, so $\vdash^{S}_{U}\Box\mathbf{A}\rightarrow\mathbf{A}$, so $\vdash^{S}_{U}\mathbf{A}$, and thus $U\mathbf{R}U$.

(A9). Let $U \in \mathbf{N}$, $U' \in \mathbf{K}$, and suppose $U\mathbf{R}U'$ and that $\vdash^{S}_{U}\mathbf{A}$. Since $U \in \mathbf{N}$, for some $\mathbf{B}$, $\vdash^{S}_{U}\Box\mathbf{B}$, and using A9, $\vdash^{S}_{U}\mathbf{A}\rightarrow\Box\mathbf{A}$, so $\vdash^{S}_{U}\Box\mathbf{A}$ and hence $\vdash^{S}_{U'}\mathbf{A}$. By the S-completeness of U and U′ it follows that $\vdash^{S}_{U}\mathbf{A}$ iff $\vdash^{S}_{U'}\mathbf{A}$ and so (identifying a theory with its set of theorems), $U = U'$.

(A10). Let $U' \in \mathbf{N}$. Then for some $\mathbf{B}$, $\vdash^{S}_{U'}\mathbf{B}$. By A10, $\vdash^{S}_{U'}\neg\Box\Box x = x$, so $\vdash^{S}_{U'}\Diamond\neg\Box x = x$. Letting $U = (d_{\neg\Box x = x}U')^{+}_{C}$, then $\vdash^{S}_{U}\neg\Box x = x$. Now if $U \in \mathbf{N}$, then for some $\mathbf{C}$, $\vdash^{S}_{U}\Box\mathbf{C}$. Now $\vdash^{S}\mathbf{C}\rightarrow x = x$, so by R2, $\vdash^{S}_{U}\Box\mathbf{C}\rightarrow\Box x = x$, and so we would have $\vdash^{S}_{U}\Box x = x$, contradicting the S-consistency of U (using Lemma 4.4). Thus $U'\mathbf{R}U \,\&\, U \in \mathbf{K} - \mathbf{N}$.

(A11). Let $U' \in \mathbf{N}$, so that for some $\mathbf{B}$, $\vdash^{S}_{U'}\Box\mathbf{B}$. Using A11, $\vdash^{S}_{U'}\Diamond\Box x \neq x$. Letting $U = (d_{\Box x \neq x}U')^{+}_{C}$, by Lemma 4.4, $U \in \mathbf{K}$ and $U'\mathbf{R}U$, and $\vdash^{S}_{U}\Box x \neq x$. Suppose $U'' \in \mathbf{K}$ and $U\mathbf{R}U''$. Then $\vdash^{S}_{U}x \neq x$, contradicting the S-consistency of U, since $\vdash^{S}x = x$. Hence there is no $U'' \in \mathbf{K}$ with $U\mathbf{R}U''$.

(A12). Let $U \in \mathbf{K}$ and suppose $U \in \mathbf{N}$. Then for some $\mathbf{B}$, $\vdash^{S}_{U}\Box\mathbf{B}$. By A12, $\vdash^{S}_{U}\Diamond x \neq x$. Then by Lemma 4.4, if $U' = (d_{x \neq x}U)^{+}_{C}$, then $U' \in \mathbf{K}$ and $U\mathbf{R}U'$. But we would have $\vdash^{S}_{U'}x \neq x$, contradicting the S-consistency of U′. Thus $\mathbf{N} = \emptyset$.

(A13). Ref_{N} follows by use of the conjunct $\Box\mathbf{A}\rightarrow\mathbf{A}$ of A13 as in (A1). Let $U, U' \in \mathbf{N}$ and suppose $U\mathbf{R}U'$. Let $\vdash^{S}_{U'}\Box\mathbf{A}$, and suppose not $-\vdash^{S}_{U}\mathbf{A}$. Hence by the S-completeness of U, $\vdash^{S}_{U}\neg\mathbf{A}$. Since by A13, $\vdash^{S}\Box\mathbf{A}\rightarrow\mathbf{A}$, we must have $\vdash^S_U \neg\Box\mathbf{A}$. Consequently by (A13), $\vdash^S_U \neg\Box\Diamond\Box\mathbf{A}$, so that $\vdash^S_U \Diamond\neg\Diamond\Box\mathbf{A}$. Since $U \in \mathbf{N}$, then $\vdash^S_U \Box\mathbf{B}$ for some $\mathbf{B}$, and hence there is a $U'' \in \mathbf{K}$ with $U\mathbf{R}U''$ such that $\vdash^S_{U''} \neg\Diamond\Box\mathbf{A}$, so $\vdash^S_{U''} \Box\neg\Box\mathbf{A}$. Hence by (A13), $\vdash^S_{U''} \neg\Box\mathbf{A}$.

(A14). To treat this general case, we first establish the following claim (cf. Lemmon and Scott (1966), Theorems 2.7 and 5.3).

For any $U, U' \in \mathbf{N}$, $U\mathbf{R}^n U'$ iff $\{\mathbf{A}: \vdash^S_U \Box^n\mathbf{A}\} \subseteq U'$. We proceed by induction on n:

$n = 0$: To show $U = U'$ iff $\{\mathbf{A}: \vdash^S_U \mathbf{A}\} \subseteq U'$. One direction is trivial. Suppose $\{\mathbf{A}: \vdash^S_U \mathbf{A}\} \subseteq U'$, $\vdash^S_{U'} \mathbf{A}$, and not $-\vdash^S_U \mathbf{A}$. By the Closure Theorem, we may assume $\mathbf{A}$ is closed. Since U is S-complete, the $\vdash^S_U \neg\mathbf{A}$, and so $\vdash^S_{U'} \neg\mathbf{A}$, contradicting the S-consistency of U. Thus $U = U'$.

For $n+1$: Assume the claim holds for n, and suppose that $U\mathbf{R}^{n+1}U'$ so that for some $U'' \in \mathbf{N}$, $U\mathbf{R}^n U''$ and $U''\mathbf{R}U'$. Suppose that $\vdash^S_U \Box^{n+1}\mathbf{A}$. Then by induction, $\vdash^S_{U''} \Box\mathbf{A}$ and so $\vdash^S_{U'} \mathbf{A}$. For the converse, suppose that $\{\mathbf{A}: \vdash^S_U \Box^{n+1}\mathbf{A}\} \subseteq U'$. Let $V = d^n(U, U')$, and suppose that V is S-inconsistent, say

$$\vdash^S \neg\mathbf{A}_1 \vee \ldots \vee \neg\mathbf{A}_k \vee \neg\Diamond^n\mathbf{B}_1 \vee \ldots \vee \neg\Diamond^n\mathbf{B}_l,$$

where for $i = 1, \ldots, k$ and $j = 1, \ldots, l$,

$$\vdash^S_U \Box\mathbf{A}_i \quad \text{and} \quad \vdash^S_{U'} \mathbf{B}_j.$$

Then

$$\vdash^S \mathbf{A}_1 \wedge \ldots \wedge \mathbf{A}_k \to \blacksquare\ \Box^n\neg\mathbf{B}_1 \vee \ldots \vee \Box^n\neg\mathbf{B}_l,$$

and so by Lemma 1.14,

$$\vdash^S \mathbf{A}_1 \wedge \ldots \wedge \mathbf{A}_k \to \Box^n[\neg\mathbf{B}_1 \vee \ldots \vee \neg\mathbf{B}_l].$$

Next, since S is a Lemmon system,

$$\vdash^S \Box[\mathbf{A}_1 \wedge \ldots \wedge \mathbf{A}_k] \to \Box^{n+1}[\neg\mathbf{B}_1 \vee \ldots \vee \neg\mathbf{B}_l],$$

and then by Lemma 1.14 again,

$$\vdash^S \Box\mathbf{A}_1 \wedge \ldots \wedge \Box\mathbf{A}_k \to \Box^{n+1}[\neg\mathbf{B}_1 \vee \ldots \vee \neg\mathbf{B}_l].$$

Thus $\vdash^S_U \Box^{n+1}[\neg\mathbf{B}_1 \vee \ldots \vee \neg\mathbf{B}_l]$, and so $\vdash^S_{U'} \neg\mathbf{B}_1 \vee \ldots \vee \neg\mathbf{B}_l$ contradicting the S-consistency of U′. Thus V must be S-consistent, and hence so is $V^+_C \in \mathbf{K}$. Since $\{\mathbf{A}: \vdash^S_U \Box\mathbf{A}\} \subseteq V \subseteq V^+_C$, then $U\mathbf{R}V^+_C$. Finally, suppose that $\vdash^S_{V^+_C} \Box^n\mathbf{B}$, but that not $\vdash^S_{U'}\mathbf{B}$. Then $\vdash^S_{U'} \neg\mathbf{B}$, so $\vdash^S_{V^+_C} \Diamond^n\neg\mathbf{B}$, and hence $\vdash^S_{V^+_C} \neg\Box^n\mathbf{B}$, contradicting the S-consistency of V^+_C. Thus $\{\mathbf{B}: \vdash^S_{V^+_C} \Box^n\mathbf{B}\} \subseteq$

U' and so by induction, $V_C^+ \mathbf{R}^n U'$. But then $U\mathbf{R}V_C^+$ and $V_C^+\mathbf{R}^n U'$, so that $U\mathbf{R}^{n+1}U'$, as desired.

Now suppose that $U_1, U_2, U_3 \in \mathbf{K}$, that $U_1\mathbf{R}^m U_2$, $U_1\mathbf{R}^p U_3$, and that for some $\mathbf{B}$, $\vdash^S_{U_1} \Box\mathbf{B}$. By our claim above,

$$\{\mathbf{A}: \vdash^S_{U_1} \Box^m \mathbf{A}\} \subseteq U_2 \quad \text{and} \quad \{\mathbf{A}: \vdash^S_{U_1} \Box^p \mathbf{A}\} \subseteq U_3.$$

Let

$$W = \{\mathbf{A}: \vdash^S_{U_2} \Box^n \mathbf{A}\} \cup \{\mathbf{B} \vdash^S_{U_3} \Box^p \mathbf{B}\}.$$

Suppose W were S-inconsistent, say

$$\vdash^S \neg \mathbf{A}_1 \vee \ldots \vee \neg \mathbf{A}_k \vee \neg \mathbf{B}_1 \vee \ldots \vee \neg \mathbf{B}_l,$$

where for $i = 1, \ldots, k$ and $j = 1, \ldots, l$, $\vdash^S_{U_2} \Box^n \mathbf{A}_i$ and $\vdash^S_{U_3} \Box^q \mathbf{B}_j$. Then

$$\vdash^S \mathbf{A}_1 \wedge \ldots \wedge \mathbf{A}_k \rightarrow \neg \mathbf{B}_1 \vee \ldots \vee \neg \mathbf{B}_l,$$

and since S is a Lemmon system,

$$\vdash^S \Box^n[\mathbf{A}_1 \wedge \ldots \wedge \mathbf{A}_k] \rightarrow \Box^n[\neg \mathbf{B}_1 \vee \ldots \vee \neg \mathbf{B}_l].$$

Using Lemma 1.14

$$\vdash^S_{U_2} \Box^n \mathbf{A}_1 \wedge \ldots \wedge \Box^n \mathbf{A}_k \rightarrow \Box^n \neg \mathbf{B}_1 \vee \ldots \vee \Box^n \neg \mathbf{B}_l,$$

and so

$$\vdash^S_{U_2} \Box^n \neg \mathbf{B}_1 \vee \ldots \vee \Box^n \neg \mathbf{B}_l.$$

Using the maximality of U_2, for some j, $\vdash^S_{U_2} \Box^n \neg \mathbf{B}_j$. Now suppose not $-\vdash^S_{U_1} \Diamond^m \Box^n \neg \mathbf{B}_j$, so that by maximality of U_1,

$$\vdash^S_{U_1} \neg \Diamond^m \Box^n \neg \mathbf{B}_j,$$

and so

$$\vdash^S_{U_2} \neg \Box^n \neg \mathbf{B}_j,$$

a contradiction. Thus $\vdash^S_{U_1} \Diamond^m \Box^n \neg \mathbf{B}_j$. But

$$\vdash^S_{U_1} \Box\mathbf{B} \rightarrow \blacksquare\ \Diamond^m \Box^n \neg \mathbf{B}_j \rightarrow \Box^p \Diamond^q \neg \mathbf{B}_j,$$

and so

$$\vdash^S_{U_1} \Box^p \Diamond^q \neg \mathbf{B}_j.$$

Consequently,

$$\vdash^S_{U_3} \Diamond^q \neg \mathbf{B}_j,$$

so that

$$\vdash^{S}_{U_3} \neg \Box^{q}\mathbf{B}_j,$$

contradicting the S-consistency of U_3. Thus W must be S-consistent. Thus $W^{+}_{C} \in \mathbf{K}$ and by our claim above, $U_2\mathbf{R}^{n}W^{+}_{C}$ and $U_3\mathbf{R}^{q}W^{+}_{C}$, as desired.

(R3). Let $U \in \mathbf{K}$. Then $\vdash^{S}_{U} x = x$, so $\vdash^{S}_{U} \Box x = x$ by R3 and so $U \in \mathbf{N}$.* ■

DEFINITION 4.12. If S is a Feys system obtained from S2 by adding the axioms $\Box Ai_1, \ldots, \Box Ai_k$ and also adding axiom pairs $\mathbf{A}_{j1}$, $\Box \mathbf{A}_{j1}$, $\mathbf{A}_{jl}$, $\Box \mathbf{A}_{jl}$, (and possibly R3), the associated Lemmon system S* is

$$S^* = C2 + \mathbf{A}1 + Ai_1 + \cdots + Ai_k + \mathbf{A}_{j1} + \cdots + \mathbf{A}_{jl}$$
$$(+\ R3 \text{ if present in S}).$$

LEMMA 4.13. If S is a Feys system, then for any $\mathbf{A}$, $\vdash^{S^*}\mathbf{A}$ implies $\vdash^{S}\mathbf{A}$.

Proof. First we show that $\vdash^{S^*}\mathbf{A}$ implies $\vdash^{S}\Box\mathbf{A}$. We proceed by induction on the length of the given proof of $\mathbf{A}$ in S*. If $\mathbf{A}$ is an axiom of S*, then $\Box\mathbf{A}$ is also an axiom of S. Now suppose that $\mathbf{A}$ is inferred from $\mathbf{B}$ and $\mathbf{B} \to \mathbf{A}$ by modus ponens. By induction we have $\vdash^{S}\Box\mathbf{B}$ and $\vdash^{S}\Box(\mathbf{B} \to \mathbf{A})$. From the latter, A0, and modus ponens, we get $\vdash^{S}\Box\mathbf{B} \to \Box\mathbf{A}$, and so by modus ponens once more, $\vdash^{S}\Box\mathbf{A}$. Finally suppose $\mathbf{A}$ is $\Box\mathbf{B} \to \Box\mathbf{C}$ and was inferred from $\mathbf{B} \to \mathbf{C}$ by R2. By induction, $\vdash^{S}\Box(\mathbf{B} \to \mathbf{C})$. Then by R2′, $\vdash^{S2}\Box(\Box\mathbf{B} \to \Box\mathbf{C})$ which is $\vdash^{S2}\Box\mathbf{A}$. But now suppose $\vdash^{S^*}\mathbf{A}$, so that $\vdash^{S}\Box\mathbf{A}$. Then by A1, $\vdash^{S}\Box\mathbf{A} \to \mathbf{A}$, so that $\vdash^{S}\mathbf{A}$. ■

DEFINITION 4.14. If S is one of the Feys systems explicitly considered in §2 and T is an S-consistent S-theory, let T* be the S*-theory whose nonlogical axioms are all $\mathbf{A}$ such that $\vdash^{S}_{T}\mathbf{A}$. Then the canonical structure for T is defined to be the canonical structure for T*.

LEMMA 4.15. Let S be one of the Feys systems explicitly considered in §2 and let T be an S-consistent S-theory. Then the canonical structure for T is an S-structure.

Proof. First note that since $\vdash^{S}_{T}\Box x = x$, then $\vdash^{S^*}_{T^*}\Box x = x$, so $\vdash^{S^*}_{U}\Box x = x$,

* Our schema A14 (m, n, p, q) corresponds to the schema G' in Section 4 of Lemmon–Scott (1966). Conditions (ii) and (iii) on $\mathbf{K}$ in Definition 4.10 are required simply to carry through the proof of Lemma 4.11 in the case corresponding to A14 – condition (i) is adequate for all of A1–A13. We can easily formulate the schema analogous to the schema H′ of Section 5 of Lemmon–Scott (1966). By adding further conditions analogous to (ii) and (iii) of Definition 4.10, the proof of Lemma 4.11 could be carried through in this case.

where $U = (T^*)_C^+$. Thus the origin **O** of the canonical structure $\mathfrak{A}$ for T belongs to **N**. Now we consider the various axioms in turn.

($\Box$A1 and A1). Let $U \in \mathbf{K}$ and suppose that $\vdash^S_U \Box \mathbf{A}$. Then by A1 (which is an axiom of every Feys system here) and modus ponens, $\vdash^S_U \mathbf{A}$, and so $U\mathbf{R}U$.

Note that by A1 and modus ponens, if $\vdash^S \Box Ai$, then $\vdash^S Ai$.

($\Box$A2)–($\Box$A7), ($\Box$A13). If $\Box Ai$ is an axiom of S, then Ai is an axiom of S*. So proceed as in the corresponding case (A2)–(A7), (A13) in the proof of Lemma 4.11.

($\Box$A15). We have $\vdash^{S^*} \Diamond\Diamond \neg \mathbf{A}$, so $\vdash^{S^*} \neg \Box\Box \mathbf{A}$. Proceed as in case (A10) of Lemma 4.11.

(R3). Proceed as in Case (R3) of Lemma 4.11. ■

LEMMA 4.16. Let S be a Lemmon system, let T be an S-consistent S-theory and let $\mathfrak{A}$ be the canonical structure for T. Then for any $k \in \mathbf{K}$ and any closed formula **A** of ML(k),

$$\vdash^S_k \mathbf{A} \quad \text{iff} \quad \mathfrak{A} \models_k \mathbf{A}.$$

Proof. We proceed by induction on the structure of **A**. First we note that is easy to prove by induction on the length of **a** that if **a** is a closed term of ML(k) with $k \in \mathbf{K}$, then $\mathbf{a}^{\mathfrak{A},k}$ is **a**. From this it immediately follows by definition that if **A** is a closed atomic formula of ML(k), then $\vdash^S_k \mathbf{A}$ iff $\mathfrak{A} \models_k \mathbf{A}$.

If **A** is $\neg \mathbf{B}$, then $\mathfrak{A} \models_k \neg \mathbf{B}$ iff not $-\mathfrak{A} \models_k \mathbf{B}$ iff (by induction) not $-\vdash^S_k \neg \mathbf{B}$ iff $\vdash^S_k \neg \mathbf{B}$, the last equivalence by virtue of the S-completeness of k. If **A** is $\mathbf{B} \vee \mathbf{C}$, then $\mathfrak{A} \models_k \mathbf{B} \vee \mathbf{C}$ iff $\mathfrak{A} \models_k \mathbf{B}$ or $\mathfrak{A} \models_k \mathbf{C}$ iff (by induction) $\vdash^S_k \mathbf{B}$ or $\vdash^S_k \mathbf{C}$ iff $\vdash^S_k \mathbf{B} \vee \mathbf{C}$, the last equivalence again by the S-completeness of k since if not $-\vdash^S_k \mathbf{B} \vee \mathbf{C}$, then $\vdash^S_k \neg(\mathbf{B} \vee \mathbf{C})$, so $\vdash^S_k \neg \mathbf{B}$ and $\vdash^S_k \neg \mathbf{C}$, which would contradict the S-consistency of k.

Now let **A** be $\exists \mathbf{x} \mathbf{B}$. If $\mathfrak{A} \models_k \exists \mathbf{x} \mathbf{B}$, then for some $\mathbf{e} \in |\mathscr{A}_k|$, $\mathfrak{A} \models_k \mathbf{B}[v(^{\mathbf{x}}_{\mathbf{e}})]$, for any v. But then since **e** is a closed term of ML(k) and $\mathbf{e}^{\mathfrak{A},k}$ is **e**, it follows that $\mathfrak{A} \models_k \mathbf{B}_x[\mathbf{e}]$. Then by induction, $\vdash^S_k \mathbf{B}_x[\mathbf{e}]$ and so by the substitution axioms (cf. Shoenfield, 1967), we have $\vdash^S_k \exists \mathbf{x} \mathbf{B}$. On the other hand, suppose that $\vdash^S_k \exists \mathbf{x} \mathbf{B}$. Since k is a Henkin Theory, there is a constant **e** such that $\vdash^S_k \exists \mathbf{x} \mathbf{B} \to \mathbf{B}_{\mathbf{x}}[\mathbf{e}]$ and so $\vdash^S_k \mathbf{B}_{\mathbf{x}}[\mathbf{e}]$. Hence by induction, $\mathfrak{A} \models_k \mathbf{B}_{\mathbf{x}}[\mathbf{e}]$. Then since $\mathbf{e}^{\mathfrak{A},k}$ is $\mathbf{e} \in |\mathscr{A}_k|$, for any v, $\mathfrak{A} \models_k \mathbf{B}[v(^{\mathbf{x}}_{\mathbf{e}})]$, and so $\mathfrak{A} \models_k \exists \mathbf{x} \mathbf{B}$.

Finally, let **A** be $\Diamond \mathbf{B}$. Suppose that $\mathfrak{A} \models_k \Diamond \mathbf{B}$. If $k \in \mathbf{K} - \mathbf{N}$, then for no **C** do we have $\vdash^S_k \Box \mathbf{C}$; in particular, not $-\vdash^S_k \Box \neg \mathbf{B}$. Hence by the S-completeness of k, $\vdash^S_k \neg \Box \neg \mathbf{B}$, and so $\vdash^S_k \Diamond \mathbf{B}$ by Theorem 1.10. If $k \in \mathbf{N}$, then for some $k' \in \mathbf{K}$ with $k\mathbf{R}k'$, we have $\mathfrak{A} \models_{k'} \mathbf{B}$. Then by induction $\vdash^S_{k'} \mathbf{B}$. Now suppose that not $-\vdash^{S}_{k} \Diamond \mathbf{B}$. Then by the S-completeness of k, $\vdash^{S}_{k} \neg \Diamond \mathbf{B}$, so again by Theorem 1.10, $\vdash^{S}_{k} \Box \neg \mathbf{B}$. Then since $k\mathbf{R}k'$, $\vdash^{S}_{k'} \neg \mathbf{B}$, contradicting the S-consistency of k'. Thus we must have $\vdash^{S}_{k} \Diamond \mathbf{B}$. For the converse, suppose that $\vdash^{S}_{k} \Diamond \mathbf{B}$. If $k \in \mathbf{K} - \mathbf{N}$, then by Definition 2.7, $\mathfrak{A} \models_{k} \Diamond \mathbf{B}$. So assume that $k \in \mathbf{N}$ so that for some $\mathbf{C}$, $\vdash^{S}_{k} \Box \mathbf{C}$. Then by Lemma 4.4 and the definition of $\mathbf{K}$, there is a $k' \in \mathbf{K}$ with $k\mathbf{R}k'$ such that $\vdash^{S}_{k'} \mathbf{B}$. Then by induction, $\mathfrak{A} \models_{k'} \mathbf{B}$, and so $\mathfrak{A} \models_{k} \Diamond \mathbf{B}$, as desired. ■

COROLLARY 4.17. Let S, T and $\mathfrak{A}$ be as in Lemma 4.16. Then $\mathfrak{A}$ is a model of T.

Proof. Let $\mathbf{A}$ be a nonlogical axiom of T and let $\mathbf{A}'$ be the universal closure of $\mathbf{A}$. Then $\vdash^{S}_{T} \mathbf{A}'$ and so $\vdash^{S}_{\mathbf{O}} \mathbf{A}'$, since $\mathbf{O} = T^{+}_{C}$. Then by Lemma 4.16, $\mathfrak{A} \models \mathbf{A}'$ and so $\mathbf{A}$ is valid in $\mathfrak{A}$. Hence $\mathfrak{A}$ is a model of T. ■

COROLLARY 4.18. Let S be a Feys system, let T be an S-consistent S-theory and let $\mathfrak{A}$ be the canonical structure for T (cf. Definition 4.14). Then $\mathfrak{A}$ is a model of T.

Proof. Let $\mathfrak{A}$ be a nonlogical axiom of T and let $\mathbf{A}'$ be the universal closure of $\mathbf{A}$. Then $\vdash^{S}_{T} \mathbf{A}'$ so that $\mathbf{A}'$ is a nonlogical axiom of T^*. Then $\vdash^{S^*}_{(T^*)^{+}_{C}} \mathbf{A}'$ and since $\mathbf{O} = (T^*)^{+}_{C}$, then by Lemma 4.16, $\mathfrak{A} \models \mathbf{A}'$. Hence $\mathbf{A}$ is valid in $\mathfrak{A}$ and so $\mathfrak{A}$ is a model of T. ■

COMPLETENESS THEOREM (4.19). Let S be one of the Lemmon or Feys systems specifically considered in §2, and let T be an S-theory. Then T is S-consistent iff there exists an S-structure $\mathfrak{A}$ which is a model of T.

Proof. If such an S-structure exists, then by the Validity Theorem, T is S-consistent. Conversely, suppose that T is S-consistent and let $\mathfrak{A}$ be the canonical structure for T. Then by Lemmas 4.11 and 4.15, $\mathfrak{A}$ is an S-structure, while by Corollaries 4.17 and 4.18, $\mathfrak{A}$ is a model of T. ■

If S is a modal system and Γ is a set of sentences of LB, we say that S is *strongly complete with respect to* Γ if any S-consistent S-theory T has a model which is a Γ-structure. We say that Γ is *strongly characteristic* for S if S is both valid and strongly complete with respect to Γ. Just as for classical logic, it is easy to see that if S is strongly complete with respect to Γ, then S is complete with respect to Γ. Our last theorem shows that if S is one of our specific Lemmon or Feys systems, then Γ_{S} is strongly characteristic (and hence characteristic) for S.

Finally, we observe that compactness follows from strong completeness. A theory T′ is a (*finitely axiomatized*) *part* of T if the (*finitely many*) *nonlogical axioms of* T′ are all nonlogical axioms of T.

COMPACTNESS THEOREM (4.20). Let Γ be strongly characteristic for the modal system S and let T be an S-theory such that for every finitely axiomatized part T′ of T, there exists a Γ-structure which is a model of T′. Then there exists a Γ-structure which is a model of T.

Proof. Suppose there were no Γ-structure which was a model of T. Since S is strongly complete with respect to Γ, then T is S-inconsistent. Hence there are nonlogical axioms $\mathbf{B}_1, \ldots, \mathbf{B}_n$ of T such that $\vdash^{S} \neg \mathbf{B}_1 \vee \ldots \vee \neg \mathbf{B}_n$. Let T′ be the S-theory whose nonlogical axioms are $\mathbf{B}_1, \ldots, \mathbf{B}_n$; then T′ is a finitely axiomatized part of T. But since S is valid with respect to Γ, T′ can have no Γ-structure as a model, contradicting our hypothesis. Therefore, there must exist a Γ-structure which is a model of T. ■

§5. LÖWENHEIM–SKOLEM THEOREMS

A *cardinal constellation* is a map c defined on a set X such that for each $x \in X$, $c(x)$ is a non-zero cardinal number. The *constellation of a modal structure* $\mathfrak{A} = \langle \mathscr{A}_k, \mathbf{K}, \mathbf{R}, \mathbf{O}, \mathbf{N} \rangle$, $c^{\mathfrak{A}}$, is that constellation c with domain $\mathbf{K}$ such that for $k \in \mathbf{K}$, $c(k) = \text{card}(|\mathscr{A}_k|)$. Let c and c' be constellations with domains X and X', respectively. We write $c \subseteq c'$ iff $X \subseteq X'$ and $c' \restriction X = c'$, and we write $c \leq c'$ iff $X = X'$ and for all $x \in X$, $c(x) \leq c'(x)$. We define card(ML) to be the cardinality of the set of all formulas of ML. A cardinal constellation c with domain including $\mathbf{K}$ is said to be *appropriate* for $\mathfrak{A}$ if for all $k, k' \in \mathbf{K}$, $k\mathbf{R}k'$ implies $c(k) \leq c(k')$ and if $\mathfrak{A}$ is a structure for ML, then $\text{card(ML)} \leq c(k)$ for all $k \in \mathbf{K}$.

DOWNWARD LÖWENHEIM–SKOLEM THEOREM (5.1). Let $\mathfrak{A} = \langle \mathscr{A}_k, \mathbf{K}, \mathbf{R}, \mathbf{O}, \mathbf{N} \rangle$ be an S-modal structure for ML and let F be a function with domain $\mathbf{K}$ such that for $k \in \mathbf{K}$, $F(k) \subseteq |\mathscr{A}_k|$. Let c be a cardinal constellation with domain $\mathbf{K}$ such that c is appropriate for $\mathfrak{A}$, $c \leq c^{\mathfrak{A}}$, and for $k \in \mathbf{K}$, $\text{card}(F(k)) \leq c(k)$ and $\text{card}\{k' : k'\mathbf{R}k\} \leq c(k)$. Then there exists an S-modal structure $\mathfrak{B} = \langle \mathscr{B}_k, \mathbf{K}, \mathbf{R}, \mathbf{O}, \mathbf{N} \rangle$ for ML such that $c^{\mathfrak{B}} = c$, for each $k \in \mathbf{K}$, $F(k) \subseteq |\mathscr{B}_k|$ and the identity map id: $\mathfrak{B} \to \mathfrak{A}$ is an elementary embedding.

Proof. We modify the construction of Tarski and Vaught (1957). For each $k \in \mathbf{K}$, let $|\mathscr{A}_k|$ be equipped with a well-ordering of type $< \text{card}(|\mathscr{A}_k|)^+$ such that if $k'\mathbf{R}k$, $k \neq k'$, $a \in |\mathscr{A}_k| - |\mathscr{A}_{k'}|$ and $b \in |\mathscr{A}_{k'}|$, then b precedes a in this well-ordering (by our hypotheses, k can have at most $c(k) \leq \text{card}(|\mathscr{A}_k|)$ predecessors). References to 'first elements of $|\mathscr{A}_k|$ such that ____' are relative to this ordering.

Now we simultaneously define the collections of sets $\{D^k_n : n < \omega\}$ for all $k \in \mathbf{K}$ by induction on n. If $n = 0$, for each $k \in \mathbf{K}$ let $D^k_0 \subseteq |\mathscr{A}_k|$ be such that $F(k) \subseteq D^k_0$ and $\text{card}(D^k_0) = c(k)$. Now assume that the D^k_m have been defined for all $k \in \mathbf{K}$ and $m \leq 2n$. Then for each $k \in \mathbf{K}$ let $D^k_{2n+1} = \cup\{D^{k'}_{2n} : k'\mathbf{R}k\}$ and let D^k_{2n+2} consist of D^k_{2n+1} together with all b such that for some formula $\mathbf{A}$ of L whose free variables are $\mathbf{x}, \mathbf{y}_1, \ldots, \mathbf{y}_p$, $p \geq 0$, and some $a_1, \ldots, a_p \in D^k_{2n+1}$, b is the first element of $|\mathscr{A}_k|$ such that $\mathfrak{A} \models_k \mathbf{A}[v(^{\mathbf{x}}_{b})]$ where $v(\mathbf{y}_i) = a_i$ for $i = 1, \ldots, p$. Clearly $m < n$ implies that $D^k_m \subseteq D^k_n$ and that $\text{card}(D^k_m) = \text{card}(D^k_n) = c(k)$ for each $k \in K$, since by induction,

$$\text{card}(D^k_{2n+1}) \leq c(k) \cdot \text{card}\{k' : k'\mathbf{R}k\} \leq c(k)^2 = c(k),$$

and

$$\text{card}(D^k_{2n+2}) \leq \text{card}(L) \cdot \text{card}(FS(D^k_{2n+1})) \leq c(k) \cdot c(k)^{\omega} = c(k),$$

where $FS(X)$ is the set of all finite sequences of elements of X.

For each k, let $|\mathscr{B}_k| = \bigcup_{n<\omega} D^k_n$, so that $F(k) \subseteq |\mathscr{B}_k|$ and $\text{card}(|\mathscr{B}_k|) = c(k)$. If $\mathbf{f}$ is an n-ary function symbol of L and $b_1, \ldots, b_n \in |\mathscr{B}_k|$, define $\mathbf{f}_{\mathscr{B}_k}(b_1, \ldots, b_n)$ to be the first $a \in |\mathscr{A}_k|$ such that $a \equiv_{\mathscr{A}_k} f_{\mathscr{A}_k}(b_1, \ldots, b_n)$. Now there must be an m such that $b_1, \ldots, b_m \in D^k_{2m}$; then the a just chosen is also the first $a \in |\mathscr{A}_k|$ such that $\mathfrak{A} \models_k \mathbf{x} = \mathbf{f}\mathbf{y}_1 \ldots \mathbf{y}_n \lfloor v(^{\mathbf{x}}_{a}) \rfloor$, where $v(\mathbf{y}_1) = b_i$ for $i = 1, \ldots, n$, and so $a \in D^k_{2m+2} \subseteq \mathbf{B}_k$. Hence $|\mathscr{B}_k|$ is closed under $\mathbf{f}_{\mathscr{B}_k}$. If $\mathbf{p}$ is any predicate symbol (including $\equiv$), let $\mathbf{p}_{\mathscr{B}_k}$ be the restriction of $\mathbf{p}_{\mathscr{A}_k}$ to $|\mathscr{B}_k|$. If $k\mathbf{R}k'$, it follows immediately from the construction that $|\mathscr{B}_k| \subseteq |\mathscr{B}_{k'}|$. It follows that there is a classical structure $\mathscr{B}_k$ for L such that $\mathscr{B}_k$ is a substructure of $\mathscr{A}_k$. This defines $\mathfrak{B}$.

Now we claim that if $\mathbf{A}$ is any formula of ML and v is any assignment in $\mathfrak{B}$, then

$$\mathfrak{B} \models_k \mathbf{A}[v] \quad \text{iff} \quad \mathfrak{A} \models_k \mathbf{A}[v].$$

We proceed by induction on the structure of $\mathbf{A}$. For atomic $\mathbf{A}$ this is immediate from the definition of substructure, while for $\mathbf{A}$ of the forms $\neg\mathbf{B}$ or $\mathbf{B} \vee \mathbf{C}$, the inductive procedure is simple. Let $\mathbf{A}$ be $\exists\mathbf{x}\mathbf{B}$ and suppose $\mathfrak{B} \models_k \exists\mathbf{x}\mathbf{B}[v]$. Then for some $b \in |\mathscr{B}_k|$, $\mathfrak{B} \models_k \mathbf{B}[v(^{\mathbf{x}}_{b})]$, so by induction, $b \in |\mathscr{A}_k|$ and $\mathfrak{A} \models_k \mathbf{B}[v(^{\mathbf{x}}_{b})]$, and hence $\mathfrak{A} \models_k \exists\mathbf{x}\mathbf{B}[v]$. On the other hand, suppose $\mathfrak{A} \models_k \exists\mathbf{x}\mathbf{B}[v]$. Then for some $a \in |\mathscr{A}_k|$, $\mathfrak{A} \models_k \mathbf{B}[v(^{\mathbf{x}}_{a})]$. Now if $\mathbf{y}$ is a free variable of $\mathbf{B}$ other than $\mathbf{x}$, $v(\mathbf{y}) \in |\mathscr{B}_k|$. Let m be such that $v(\mathbf{y}) \in D^k_{2m+1}$ for each free $\mathbf{y}$ in $\mathbf{B}$ other than $\mathbf{x}$, and let b be the first element of $\mathscr{A}_k$ such that $\mathfrak{A} \models_k \mathbf{B}[v(^{\mathbf{x}}_{b})]$. Then $b \in D^k_{2m+2} \subseteq |\mathscr{B}_k|$, so $v(^{\mathbf{x}}_{b})$ is an assignment in $\mathfrak{B}$, hence by induction $\mathfrak{A} \models_k \mathbf{B}[v(^{\mathbf{x}}_{b})]$ and therefore, $\mathfrak{B} \models_k \exists\mathbf{x}\mathbf{B}[v]$.

Finally, suppose that $\mathbf{A}$ is $\Diamond\mathbf{B}$. Then since $b(\mathfrak{B}) = b(\mathfrak{A})$, using the induction hypothesis we have: $\mathfrak{A} \models_k \Diamond\mathbf{B}[v]$ iff $\exists k'(k\mathbf{R}k' \,\&\, \mathfrak{A} \models_{k'} \mathbf{B}[v])$ iff $\exists k'(k\mathbf{R}k' \,\&\, \mathfrak{B} \models_{k'} \mathbf{B}[v])$ iff $\mathfrak{B} \models_k \Diamond\mathbf{B}[v]$. Thus $\mathfrak{B} \prec \mathfrak{A}$, as desired. ■

The following more defined version of this theorem will be necessary below.

THEOREM 5.2. Let $\mathfrak{A} = \langle \mathscr{A}_k, \mathbf{K}, \mathbf{R}, \mathbf{O}, \mathbf{N} \rangle$ be a modal structure for ML, let $\mathbf{O} \in J \subseteq \mathbf{K}$, let c be a constellation with domain J, let F be a function with domain J, let $\bar{\mathbf{R}}$ be the least transitive reflexive relation with domain $\mathbf{K}$ containing $\mathbf{R}$, and assume the following:

(i) for $k \in J$, $\emptyset \neq F(k) \subseteq |\mathscr{A}_k|$.
(ii) for $k \in J$, $\mathrm{card}(F(k)) \leq c(k) \leq \mathrm{card}|\mathscr{A}_k|$ and $\mathrm{card}(\mathrm{ML}) \leq c(k)$
(iii) for $k, l \in J$, $k\bar{\mathbf{R}}l$ implies $c(k) \leq c(l)$.
(iv) for $k \in J, \mathrm{card}(\{k' \in J : k'\bar{\mathbf{R}}k\}) \leq c(k)$.

Then there exists a modal structure $\mathfrak{B} = \langle \mathscr{B}_l, \mathbf{L}, \mathbf{S}, \mathbf{P}, \mathbf{M} \rangle$ such that:

(1°) $J \subseteq \mathbf{L} \subseteq \mathbf{K}$, $\mathbf{O} = \mathbf{P}$, and $\mathbf{R} \upharpoonright \mathbf{L}^2 = \mathbf{S}$.
(2°) for $l \in \mathbf{L}$, $\mathscr{B}_l \subseteq \mathscr{A}_l$.
(3°) for $l \in J$, $F(l) \subseteq |\mathscr{B}_l|$ and $\mathrm{card}(|\mathscr{B}_l|) = c(l)$.
(4°) the identity map id: $\mathfrak{B} \to \mathfrak{A}$ is an elementary embedding.

Proof. For each $k \in J$, let $|\mathscr{A}_k|$ be equipped with a well-ordering of type $<\mathrm{card}(|\mathscr{A}_k|)^+$ such that if $k, k' \in J$, $k'\mathbf{R}k$, $k' \neq k$, $a \in |\mathscr{A}_k| - |\mathscr{A}_{k'}|$, and $b \in |\mathscr{A}_{k'}|$, then b occurs before a in this ordering (by hypothesis (iv), k has at most $c(k) \leq \mathrm{card}(|\mathscr{A}_k|)$ $\bar{\mathbf{R}}$-predecessors in J, and by hypotheses (ii) and (iii), for each such predecessor, there are at most $\mathrm{card}(|\mathscr{A}_k|)$ candidates for b). Reference to 'first elements of $|\mathscr{A}_k|$ such that ...' will be relative to this ordering. Let **K** be well-ordered in type card(**K**). References to 'first elements of **K** such that...' will be relative to this ordering.

We will define sets J_n and D_n^k for $k \in J_n$ and also extensions c_n of c to J_n, simultaneously by induction on n.

Stage $n = 0$: Set $J_0 = J$, $c_0 = c$, and for $k \in J_0$, let $D_0^k \subseteq |\mathscr{A}_k|$ be such that $\mathrm{card}(D_0^k) = c_0(k)$ and $F(k) \subseteq D_0^k$.

Now we will write $X_{(n)} = \bigcup_{m<n} X_m$, $X''k = \{l \in \mathbf{K} : lXk\}$, and $D_n^{*k} = \cup\{D_{(n)}^{k'} : k' \in J_{(n)} \cap \mathbf{R}''k\}$.

Stage $n > 0$:

(i) $n \equiv 0 (\mathrm{mod}\ 3)$. Set $J_n = J_{(n)}$ and $c_n = c_{n-1}$. For each $k \in J_n$, set $D_n^k = D_n^{*k}$.

(ii) $n \equiv 1 (\mathrm{mod}\ 3)$. For $k \in J_{(n)}$, let $\mathrm{ext}_n(k)$ be the set of all $l \in \mathbf{K}$ such that for some formula $\Diamond\mathbf{A}$ of ML with free variables $\mathbf{y}_1, \ldots, \mathbf{y}_n$, $p \geq 0$, and for some $a_1, \ldots, a_p$, $p \geq 0$, in D_n^{*k}, l is the first element of **K** such that $k\mathbf{R}l$ and $\mathfrak{A} \models_l \mathbf{A}[v]$, where $v(\mathbf{y}_i) = a_i$ for $i = 1, \ldots, p$. Then set $J_n = J_{(n)} \cup \cup\{\mathrm{ext}_n(k): k \in J_{(n)}\}$. For $k \in J_{(n)}$, set $D_n^k = D_n^{*k}$, and set $c_n(k) = c_{n-1}(k)$. For $l \in J_n - J_{(n)}$, let $D_0^l = \cdots = D_n^l \subseteq A_1$ where D_n^l is such that $\cup\{D_n^k : k \in J_{(n)} \cap \bar{\mathbf{R}}''l\} \subseteq D_n^l$ and $\mathrm{card}(D_n^l) \geq \mathrm{card}(J_{(n)} \cap \bar{\mathbf{R}}''l)$. Set $c_n(l) = \mathrm{card}(D_n^l)$; moreover, let $\mathrm{card}(D_n^l)$ be as small as possible within the foregoing restrictions.

(iii) $n \equiv 2 (\mathrm{mod}\ 3)$. Set $J_n = J_{(n)}$, $c_n = c_{n-1}$. For $k \in J_n$, let D_n^k consist of $D_{(n)}^k$ together with all $b \in |\mathscr{A}_k|$ such that for some formula **A** of ML with free variables $\mathbf{x}, \mathbf{y}_1, \ldots, \mathbf{y}_p$, $p \geq 0$, and for some $a_1, \ldots, a_p \in D_n^{*k}$, b is the first element of $|\mathscr{A}_k|$ such that $\mathfrak{A} \models_k \mathbf{A}[v(^{\mathbf{x}}_b)]$, $v(\mathbf{y}_i) = a_i$, $i = 1, \ldots, p$.

We clearly have: $m \leq n$ implies $J_m \subseteq J_n$ and $D_m^k \subseteq D_n^k$. Then set $\mathbf{L} = \bigcup_{n<\omega} J_n$ and for $k \in \mathbf{L}$, set $|\mathscr{B}_k| = \bigcup_{n<\omega} D_n^k$. Since step (i) in the above construction is repeated infinitely often, it is clear that if k, $l \in \mathbf{L}$ and $k\bar{\mathbf{R}}l$, then $|\mathscr{B}_k| \subseteq |\mathscr{B}_l|$.

Next we claim that for any $m \leq n$ and any $k \in J_m$, the following all hold:

$$c_n(k) = c_m(k), \quad \operatorname{card}(D_n^k) = c_n(k), \quad \operatorname{card}(J_n \cap \bar{\mathbf{R}}''k) \leq c_n(k),$$

$$\text{and} \quad k, k' \in J_n \quad \text{and} \quad k'\bar{\mathbf{R}}k \quad \text{imply that} \quad c_n(k') \leq c_n(k).$$

We verify this claim by induction on n. Under the hypotheses of the theorem, the statements are immediate for $n = 0$. So assume that $n > 0$ and $0 \leq m < n$.

(i) $n \equiv 0 (\operatorname{mod} 3)$. Since $c_n = c_{n-1}$, $c_n(k) = c_m(k)$ is immediate by induction. Now $J_n = J_{(n)}$, so $J_n \cap \bar{\mathbf{R}}''k = \bigcup_{q<n} J_q \cap \bar{\mathbf{R}}''k$ and so by induction, $\operatorname{card}(J_n \cap \bar{\mathbf{R}}''k) \leq \sum_{q<n} \operatorname{card}(J_q \cap \bar{\mathbf{R}}''k) \leq \sum_{q<n} c_q(k) = n \cdot c_n(k) = c_n(k)$. Thus using induction,

$$c_n(k) = c_m(k) \leq \operatorname{card}(D_n^k) \leq \sum_{k' \in J_n \cap \bar{\mathbf{R}}''k} \sum_{q<n} c_q(k')$$
$$\leq \sum_{k' \in J_n \cap \bar{\mathbf{R}}''k} c_n(k) \leq c_n(k)^2 = c_n(k).$$

Since $J_n = J_{(n)}$ and $c_n(k) = c_{n-1}(k)$, the last statement of the claim follows by induction.

(ii) $n \equiv 1 (\operatorname{mod} 3)$. For $k \in J_m$, $c_n(k) = c_{n-1}(k)$ and so $c_n(k) = c_m(k)$ follows. For $k \in J_{(n)}$, $\operatorname{card}(D_n^k) = c_n(k)$ by the same argument as given for (i) above since D_n^{*k} only depends on $J_{(n)}$. And for $l \in J_n - J_{(n)}$, $c_n(l) = \operatorname{card}(D_n^l)$ by definition. Now let $k \in J_{(n)}$. Using the facts above and induction, we have:

$$\operatorname{card}(\operatorname{ext}_n(k)) \leq \operatorname{card}(\mathrm{ML}) \cdot \sum_{q<n} \sum_{k' \in J_q \cap \mathbf{R}''k} \sum_{s<n} c_s(k')$$
$$\leq c_n(k) \cdot \sum_{q<n} \sum_{k' \in J_n \cap \bar{\mathbf{R}}''k} c_q(k')$$
$$\leq c_n(k) \cdot \sum_{q<n} \sum_{k' \in J_q \cap \bar{\mathbf{R}}''k} c_q(k)$$
$$\leq c_n(k) \cdot \sum_{q<n} c_q(k)^2 = c_n(k).$$

Again let $k \in J_{(n)}$ and suppose that $k' \in J_{(n)}$ and $l \in \operatorname{ext}_n(k') \cap \bar{\mathbf{R}}''k$. Then $k'\bar{\mathbf{R}}l$ and $l\bar{\mathbf{R}}k$, so $k'\bar{\mathbf{R}}k$. Thus by induction,

$$\operatorname{card}(J_n \cap \bar{\mathbf{R}}''k) \leq \operatorname{card}(J_{(n)} \cap \bar{\mathbf{R}}''k) + \sum_{k' \in J_{(n)}} \operatorname{card}(\operatorname{ext}_n(k') \cap \bar{\mathbf{R}}''k)$$
$$\leq c_n(k) + \sum_{q<n} \sum_{k' \in J_q \cap \bar{\mathbf{R}}''k} \operatorname{card}(\operatorname{ext}_n(k') \cap \bar{\mathbf{R}}''k)$$

$$\le c_n(k) + \sum_{q<n} \sum_{k' \in J_q \cap \bar{\mathbf{R}}''k} c_n(k')$$
$$\le c_n(k) + \sum_{q<n} \sum_{k' \in J_q \cap \bar{\mathbf{R}}''k} c_n(k)$$
$$\le c_n(k) + \sum_{q<n} c_q(k) \cdot c_n(k) = c_n(k).$$

And if $l \in J_n - J_{(n)}$, then $c_n(l) = \text{card}(D_n^l) \ge \text{card}(J_{(n)} \cap \bar{\mathbf{R}}''l)$. Suppose that $l' \in J_n - J_{(n)}$ and $l'\bar{\mathbf{R}}l$. Then for some $k' \in J_{(n)}$, $l' \in \text{ext}_n(k)$, so $k'\mathbf{R}l'$ and $l'\bar{\mathbf{R}}l$, so $k'\bar{\mathbf{R}}l$. Thus $(J_n - J_{(n)}) \cap \bar{\mathbf{R}}''l \bigcup_{k' \in J_{(n)} \cap \mathbf{R}''l} \text{ext}_n(k') \cap \bar{\mathbf{R}}''l$. Hence

$$\text{card}(J_n \cap \bar{\mathbf{R}}''l) \le c_n(l) + \sum_{k' \in J_{(n)} \cap \bar{\mathbf{R}}''l} c_n(k')$$
$$\le c_n(l) + \sum_{k' \in J_{(n)} \cap \bar{\mathbf{R}}''l} c_n(l) \le c_n(l) + c_n(l)^2 = c_n(l).$$

Finally, let $k', k \in J_n$ and suppose that $k'\bar{\mathbf{R}}k$. Then:

(a) If $k, k' \in J_{(n)}$, induction implies that $c_n(k') \le c_n(k)$.

(b) If $k' \in J_{(n)}$ and $k \in J_n - J_{(n)}$, then by induction and the definition of D_n^k, $c_n(k') = \text{card}(D_n^{k'}) \le \text{card}(D_n^k) = c_n(k)$.

(c) Let $k' \in J_n - J_{(n)}$ and $k \in J_{(n)}$. Then

$$c_n(k') = \max(\text{card}(J_{(n)} \cap \bar{\mathbf{R}}''k'), \sup_{k'' \in J_{(n)} \cap \bar{\mathbf{R}}''k'} c_n(k')).$$

Let $k'' \in J_{(n)} \cap \bar{\mathbf{R}}''k'$, so that $k''\bar{\mathbf{R}}k'$. Since $k'\bar{\mathbf{R}}k'$, then $k''\bar{\mathbf{R}}k$. Since $k'', k' \in J_{(n)}$, by induction we have $c_n(k'') \le c_n(k)$. Hence $c_n(k') \le c_n(k)$.

(d) Let $k, k' \in J_n - J_{(n)}$. Then by the transitivity of $\bar{\mathbf{R}}$, $J_{(n)} \cap \bar{\mathbf{R}}''k' \subseteq J_{(n)} \cap \bar{\mathbf{R}}''k$ and the claim follows.

(iii) $n \equiv 2 (\text{mod } 3)$. Then $J_n = J_{(n)}$ and $c_n = c_{n-1}$, so $c_m(k) = c_n(k)$, $\text{card}(J_n \cap \bar{\mathbf{R}}''k) \le c_n(k)$, and $k, k' \in J_n \& k'\bar{\mathbf{R}}k$ implies $c_n(k') \le c_n(k)$. And:

$$c_n(k) = c_{n-1}(k) \le \text{card}(D_{(n)}^k) + \text{card}(\text{ML}) \cdot \text{card}(\text{FS}(D_n^{*k}))$$
$$\le \sum_{q<n} c_q(k) + c_n(k) \cdot \sum_{j=0}^{\omega} (\text{card}(D_n^{*k}))^j$$
$$\le \sum_{q<n} c_n(k) + c_n(k) \cdot \aleph_0 \cdot \text{card}(D_n^{*k}) = \text{card}(D_n^{*k})$$
$$\le \sum_{k' \in J_{(n)} \cap \bar{\mathbf{R}}''k} c_n(k') \le \sum_{k' \in J_{(n)} \cap \bar{\mathbf{R}}''k} c_n(k)$$
$$= c_n(k) \cdot \text{card}(J_{(n)} \cap \bar{\mathbf{R}}''k) \le c_n(k)^2 = c_n(k).$$

This verifies the claim. Setting $\mathbf{S} = \mathbf{R} \restriction \mathbf{L}^2$ and for $k \in \mathbf{L}$, $\mathscr{B}_k = \mathscr{A}_k \restriction |\mathscr{B}_k|$, the theorem now follows. ■

If $\mathfrak{A}$ is a modal structure, we define the *cardinality of* $\mathfrak{A}$ to be $\text{Card}(\mathfrak{A}) = \text{Card}(\text{sk}(\mathfrak{A}))$. The next theorem presents a simple version of the upward Löwenheim–Skolem phenomena.

THEOREM 5.3. Let S be a Lemmon or Feys system which is strongly complete with respect to a set Γ and let T be an S-consistent S-theory which possesses an S-model $\mathfrak{B}$ in which $|\mathscr{B}_{\mathbf{0}}|$ is infinite. Then for each $\kappa \geq \max(\aleph_0, \mathrm{card}(\mathrm{ML}))$ there exists a Γ-structure $\mathfrak{A}$ which is a model of T such that

$$\mathrm{card}(\mathfrak{A}) = \mathrm{card}(|\mathscr{A}_{\mathbf{0}}|) = \kappa.$$

Proof. Let $\{c_\delta : \delta < \kappa\}$ be a set of mutually distinct individual constants none of which occurs in the language ML(T). Let T′ be the theory whose axioms are those of T together with $\{c_\delta \neq c_\gamma : \delta < \gamma < \kappa\}$. Clearly the given model $\mathfrak{B}$ can be expanded to a model of T″ for any finitely axiomatized part T″ of T′. Hence by the Compactness Theorem, there is a Γ-structure which is a model of T′, so that T′ is S-consistent. Let $\mathfrak{A}$ be the canonical structure for T′. By Corollaries 4.17 and 4.18, $\mathfrak{A}$ is a model of T′ and hence of T. Now note that $\mathrm{Card}(\mathrm{ML}(\mathrm{T}')) = \kappa$. Then since **K** can be generated (cf. Definition 4.10) in a countable sequence of steps in which at most κ elements are added at each step, then $\mathrm{Card}(\mathbf{K}) \leq \kappa$. From the definition of Henkin extension, we always have $\mathrm{Card}(\mathrm{T}_c) = \mathrm{Card}(\mathrm{T})$. Since T′ contains exactly κ individual constants, it follows that for every $k \in \mathbf{K}$, $\mathrm{Card}(|\mathscr{A}_k|) = \kappa$. Hence

$$\kappa \leq \mathrm{Card}(\mathrm{sk}(\mathscr{A})) \leq \mathrm{Card}(\mathbf{K}) \cdot (1 + \kappa) = \kappa + \kappa^2 = \kappa. \; \blacksquare$$

To establish a more refined version of the upward theorem, we make use of a translation from modal to classical formulas suggested by Haspel (1972) and then exploit the resulting correspondence between modal and classical structures. Let S be a modal system and let T be an S-theory. We make use of a two-sorted classical language LS_T which contains the one-sorted classical language LB (cf. Kreisel and Krivine (1967) and Wang (1952) or (1970), Chapter XII). The primitive symbols of LS_T are:

variables sort 1: $w, w', w'', \ldots$

sort 2: $x \; y \; z \; x' \; y' \; z' \; x'' \ldots$;

for each n-ary function symbol **f** of ML(T), an $(n+1)$-ary function symbol $\mathbf{f}^*$ for all $n \geq 0$;

for each n-ary predicate symbol **p** (including equality) of ML(T), an $(n+1)$-ary predicate symbol $\mathbf{p}^*$ for all $n \geq 0$;

equality: =

unary constant: N*

binary constants: R* and B*

individual constant: O*

logical symbols: $\neg \quad \vee \quad \exists$.

The formation rules for *terms* and *atomic formulas* are as follows:

(i) any variable of sort i is a term of sort i for $i = 1, 2$.

(ii) O^* is a term of sort 1.

(iii) if $\mathbf{f}$ is an n-ary function symbol of ML(T), if $\mathbf{a}$ is a term of sort 1 and $\mathbf{b}_1, \ldots, \mathbf{b}_n$ are terms of sort 2, then $\mathbf{f^*ab}_1 \ldots \mathbf{b_n}$ is a term of sort 2.

(iv) if $\mathbf{a}$ and $\mathbf{b}$ are both terms of sort i, then $=\mathbf{ab}$ is an atomic formula, $i = 1, 2$.

(v) if $\mathbf{a}$, $\mathbf{c}$ are terms of sort 1 and $\mathbf{b}$ is a term of sort 2, then the following are atomic formulas: $\mathrm{N}^*\mathbf{a}$, $\mathbf{a}\mathrm{R}^*\mathbf{c}$, and $\mathbf{a}\mathrm{B}^*\mathbf{b}$.

(vi) if $\mathbf{p}$ is an n-ary predicate symbol of ML(T), if $\mathbf{a}$ is a term of sort 1 and $\mathbf{b}_1, \ldots, \mathbf{b}_n$ are terms of sort 2, then $\mathbf{p^*ab}_1 \ldots \mathbf{b_n}$ is an atomic formula.

Formulas are built up from atomic formulas using $\neg$, $\vee$, and $\exists$ in the usual way, where quantification is allowed on both sorts of variables.

DEFINITION 5.4. Given an arbitrary term $\mathbf{a}$ of ML(T), we define the *translation* $\mathbf{a}^*$ *of* $\mathbf{a}$ *in* $\mathrm{LS_T}$ in such a way that $\mathbf{a}^*$ contains at most one variable of sort 1. The definition is by induction as follows:

$(\mathbf{x})^*$ is $\mathbf{x}$.

$(\mathbf{fa}_1 \ldots \mathbf{a}_n)^*$ is $\mathbf{f^*wa}_1^\# \ldots \mathbf{a}_n^\#$, where $\mathbf{w}$ is the first variable of sort 1 not occurring in $\mathbf{a}_1^*, \ldots, \mathbf{a}_n^*$, and $\mathbf{a}_i^\#$ is the result of substituting $\mathbf{w}$ for the free variable of sort 1 occurring in $\mathbf{a}_i^*$, if any, for $i = 1, \ldots, n$.

DEFINITION 5.5. Given a formula $\mathbf{A}$ of ML(T), we define a formula $\mathbf{A}^*$ of $\mathrm{LS_T}$ in such a way that $\mathbf{A}^*$ contains exactly one free variable of sort 1. The definition is by induction as follows:

$(=\mathbf{ab})^*$ is $=^*\mathbf{wa}^\#\mathbf{b}^\#$, where $\mathbf{w}$ is the first variable of sort 1 not occurring in either $\mathbf{a}^*$ or $\mathbf{b}^*$, and $\mathbf{a}^\#$ and $\mathbf{b}^\#$ are the results respectively of substituting $\mathbf{w}$ for the free variable of $\mathbf{a}^*$ and $\mathbf{b}^*$, if any.

$(\mathbf{pa}_1 \ldots \mathbf{a_n})^*$ is $\mathbf{p^*wa}_1^\# \ldots \mathbf{a_n}^\#$, where $\mathbf{w}$ is the first variable of sort 1 not occurring in any of $\mathbf{a}_1^*, \ldots, \mathbf{a}_n^*$, and for $i = 1, \ldots, n$, $\mathbf{a}_i^\#$ is the result of substituting $\mathbf{w}$ for the free variable of $\mathbf{a}_i^*$, if any.

$(\neg\mathbf{A})^*$ is $\neg\mathbf{A}^*$.

$(\vee\mathbf{AB})^*$ is $\vee\mathbf{A}^\#\mathbf{B}^\#$, where if $\mathbf{w}$ is the first variable of sort 1 not occurring in $\mathbf{A}^*$ or $\mathbf{B}^*$, then $\mathbf{A}^\#$ (resp., $\mathbf{B}^\#$) is the result of substituting w for the free variable of sort 1 in $\mathbf{A}^*$ (resp. $\mathbf{B}^*$).

$(\exists\mathbf{x}\mathbf{A})^*$ is $\exists\mathbf{x}[\mathbf{w}\mathbf{B}^*x \,\&\, \mathbf{A}^*]$, where $\mathbf{w}$ is the free variable of sort 1 occurring in $\mathbf{A}^*$.

$(\Diamond\mathbf{A})^*$ is $\neg\mathbf{N}^*\mathbf{w}' \vee \exists\mathbf{w}[\mathbf{w}'\mathrm{R}^*\mathbf{w} \,\&\, \mathbf{A}^*]$, where $\mathbf{w}$ is the free variable of sort 1 occurring in $\mathbf{A}^*$ and $\mathbf{w}'$ is the first variable of sort 1 not occurring in $\mathbf{A}^*$.

DEFINITION 5.6. Given a modal structure $\mathfrak{A}$ for ML(T), we define a classical two-sorted structure $\mathfrak{A}^* = \langle \mathrm{W}, \mathrm{I}, \ldots \rangle$ for $\mathrm{LS_T}$ (cf. Kreisel and Krivine, 1967) as follows. Let $\mathfrak{A} = \langle \mathscr{A}_k, \mathbf{K}, \mathbf{R}, \mathbf{O}, \mathbf{N} \rangle$. Then $\mathrm{W} = \mathbf{K}$, $\mathrm{I} = \mathrm{U}(\mathfrak{A}) = \bigcup_{k\in\mathbf{K}} |\mathscr{A}_k|$, $\mathrm{O}^*_{\mathfrak{A}^*}$ is $\mathbf{O}$, $\mathrm{N}^*_{\mathfrak{A}^*}$ is $\mathbf{N}$, and $\mathrm{R}^*_{\mathfrak{A}^*}$ is $\mathbf{R}$. For $\alpha\in\mathrm{W}$ and $\beta\in\mathrm{I}$ we define

$$\langle \alpha, \beta \rangle \in \mathrm{B}^*_{\mathfrak{A}^*} \quad \text{iff} \quad \beta \in |\mathscr{A}_\alpha|.$$

If $\mathbf{f}$ is an n-ary function symbol of ML(T), $\alpha\in\mathrm{W}$, and $\beta_1, \ldots, \beta_n \in \mathrm{I}$, we set

$$\mathbf{f}^*_{\mathfrak{A}^*}(\alpha, \beta_1, \ldots, \beta_n) = \begin{cases} \mathbf{f}_{\mathscr{A}_\alpha}(\beta_1, \ldots, \beta_n) \text{ if } \beta_1, \ldots, \beta_n \in |\mathscr{A}_\alpha| \\ \beta_i \text{ otherwise, where } i \text{ is minimal with } \beta_i \notin |\mathscr{A}_\alpha|, \end{cases}$$

and if $\mathbf{p}$ is an n-ary predicate symbol of ML(T), $\alpha\in\mathrm{W}$, and $\beta_1, \ldots, \beta_n \in \mathrm{I}$, we define

$$\mathbf{p}^*_{\mathfrak{A}^*}(\alpha, \beta_1, \ldots, \beta_n) \quad \text{iff} \quad \mathbf{p}_{\mathscr{A}_\alpha}(\beta_1, \ldots, \beta_n).$$

This specifies $\mathfrak{A}^*$.

An *assignment* in a structure $\langle \mathrm{W}, \mathrm{I}, \ldots \rangle$ for $\mathrm{LS_T}$ is a map μ defined on all the variables of $\mathrm{LS_T}$ such that if $\mathbf{w}$ is a variable of sort 1, then $\mu(\mathbf{w}) \in \mathrm{W}$ and if $\mathbf{x}$ is a variable of sort 2, then $\mu(\mathbf{x}) \in \mathrm{I}$. Let $\mathfrak{A}$ be a modal structure for ML(T), let $k\in\mathbf{K}$, and let v be as assignment in $\mathfrak{A}$. Then v^*_k is that assignment in $\mathfrak{A}^*$ defined by: $v^*_k(\mathbf{w}) = k$ if $\mathbf{w}$ is a variable of sort 1, and $v^*_k(\mathbf{x}) = v(\mathbf{x})$ if $\mathbf{x}$ is a variable of sort 2.

THEOREM 5.7. Let $\mathfrak{A}$ be a modal structure for ML(T), let $k\in\mathbf{K}$, and let v be an assignment in $\mathfrak{A}$. Then for any term $\mathbf{a}$ of ML(T), we have

$$\mathbf{a}^{\mathfrak{A},k}[v] = (\mathbf{a}^*)^{\mathfrak{A}^*}[v^*_k],$$

and for any formula $\mathbf{A}$ of ML(T),

$$\mathfrak{A} \models_k \mathbf{A}[v] \quad \text{iff} \quad \mathfrak{A}^* \models \mathbf{A}^*[v^*_k].$$

Proof. We proceed by induction on the lengths of $\mathbf{a}$ and $\mathbf{A}$. For terms, the basis step is obvious and the induction step follows immediately from the definition of $\mathbf{f}^*_{\mathfrak{A}^*}$. For formulas, the result is immediate for atomic **A** from the definitions and for **A** of the form $\neg\mathbf{B}$, $\mathbf{B} \vee \mathbf{C}$, or $\exists\mathbf{x}\mathbf{B}$, the induction is simple, the last case making use of the definition of $\mathrm{B}^*_{\mathfrak{A}^*}$. Suppose **A** is $\Diamond\mathbf{B}$ so that $\mathbf{A}^*$ is $\neg \mathrm{N}^*w' \vee \exists\mathbf{w}[\mathbf{w}'\mathrm{R}^*\mathbf{w} \,\&\, \mathbf{B}^*]$, where **w** is the free variable of sort 1 occurring in $\mathbf{B}^*$ and $\mathbf{w}'$ is the first variable of sort 1 not occurring in $\mathbf{B}^*$. Suppose that $\mathfrak{A}\models_k\mathbf{A}[v]$. If $k\in\mathbf{K}-\mathbf{N}$, then regarding k as an element of the domain W of $\mathfrak{A}^*$, we have not $-\mathrm{N}^*_{\mathfrak{A}^*}(k)$, and so $\mathfrak{A}^*\models\mathbf{A}^*[v^*_k]$. If $k\in\mathbf{N}$, then for some $k'\in\mathbf{K}$ with $k\mathbf{R}k'$, $\mathfrak{A}\models_{k'}\mathbf{B}[v]$. Then $k\mathrm{R}^*_{\mathfrak{A}^*}k'$ and by induction, $\mathfrak{A}^*\models\mathbf{B}^*[v^*_{k'}]$. But $\mathfrak{A}^*\models\mathbf{B}^*[v^*_k]$ iff $\mathfrak{A}^*\models\mathbf{B}^*[v^*_k(^{\mathbf{w}}_{k'})]$, so that $\mathfrak{A}^*\models\mathbf{w}'\mathrm{R}^*\mathbf{w} \,\&\, \mathbf{B}^*[v^*_k(^{\mathbf{w}}_{k'})]$, and hence $\mathfrak{A}^*\models\exists\mathbf{w}[\mathbf{w}'\mathrm{R}^*\mathbf{w} \,\&\, \mathbf{B}^*][v^*_k]$, as desired. Conversely, suppose that $\mathfrak{A}^*\models\mathbf{A}^*[v^*_k]$. Then either $\mathfrak{A}^*\models\neg\,\mathbf{N}^*(\mathbf{w}')[v^*_k]$ or $\mathfrak{A}^*\models\exists\mathbf{w}[\mathbf{w}'\mathrm{R}^*\mathbf{w} \,\&\, \mathbf{B}^*][v^*_k]$. In the former case, not $-\mathrm{N}^*_{\mathfrak{A}^*}(k)$, so $k\in\mathbf{K}-\mathrm{N}$, and hence $\mathfrak{A}\models_k\Diamond\mathbf{B}[v]$. In the latter case, for some $k'\in\mathrm{W}$, $\mathfrak{A}^*\models\mathbf{w}'\mathrm{R}^*\mathbf{w} \,\&\, \mathbf{B}^*[v^*_k(^{\mathbf{w}}_{k'})]$. Then $k\mathrm{R}^*_{\mathfrak{A}^*}k'$, so $k\mathbf{R}k'$. Moreover, $\mathfrak{A}^*\models\mathrm{B}^*[v^*_k(^{\mathbf{w}}_{k'})]$, so $\mathfrak{A}^*\models\mathbf{B}^*[v^*_{k'}]$. Then by induction, $\mathfrak{A}\models_{k'}\mathbf{B}[v]$. It follows that $\mathfrak{A}\models_k\Diamond\mathbf{B}[v]$, as desired. ■

We say that a modal structure $\mathfrak{A} = \langle \mathscr{A}_k, \mathbf{K}, \ldots \rangle$ has a *base of power* κ if $\mathrm{card}(\mathbf{K}) = \kappa$.

LEMMA 5.8. Let $\mathfrak{A} = \langle \mathscr{A}_k, \mathbf{K}, \ldots \rangle$ be a modal structure. Then for any $\kappa \geq \mathrm{card}(\mathbf{K})$, there exists an elementary extension $\mathfrak{B}$ of $\mathfrak{A}$ with a base of power κ.

Proof. Let $\{p_\gamma : \gamma < \kappa\}$ be a set of mutually distinct entities, none of which occur in $\mathrm{sk}(\mathfrak{A})$. Define $\mathfrak{B} = \langle \mathscr{B}_l, \mathbf{L}, \mathbf{S}, \mathbf{P}, \mathbf{M} \rangle$ as follows. Let $\mathbf{L} = \mathbf{K} \cup \{p_\gamma : \gamma < \kappa\}$ and set $\mathbf{P} = \mathbf{O}$. Define **S** so that $\mathrm{S}\restriction\mathbf{K} = \mathbf{R}$, and for all $\gamma < \kappa$ and $k\in\mathbf{K}$,

$$k\mathbf{S}p_\gamma \text{ iff } k\mathbf{RO} \text{ and } p_\gamma\mathbf{S}k \text{ iff } \mathbf{OR}k, \text{ and for } \delta \neq \gamma,\ p_\delta\mathbf{S}p_\gamma \text{ iff } \mathbf{ORO}.$$

Define **M** so that $\mathbf{M}\restriction\mathbf{K} = \mathbf{N}$ and $\mathbf{M}(p_\gamma)$ iff $\mathbf{N}(\mathbf{O})$. Finally, for $l\in\mathbf{L}$, set $\mathscr{B}_l = \mathscr{A}_l$ if $l\in\mathbf{K}$ and $\mathscr{B}_l = \mathscr{A}_{\mathbf{O}}$ if $l = p_\gamma$. We claim that for any $k\in\mathbf{K}$, assignment v in $\mathfrak{A}$, formula **A**, and $\gamma < \kappa$,

$$\mathfrak{A}\models_k\mathbf{A}[v] \quad \text{iff} \quad \mathfrak{B}\models_k\mathbf{A}[v],$$

and

$$\mathfrak{B}\models_{\mathbf{p}}\mathbf{A}[v] \quad \text{iff} \quad \mathfrak{B}\models_{\mathbf{p}_\gamma}\mathbf{A}[v].$$

We proceed by induction on the structure of **A**. For atomic **A** this follows immediately from the definition of $\mathfrak{B}$ while **A** of the form $\neg\mathbf{B}$, $\mathbf{B}\vee\mathbf{C}$, or $\exists\mathbf{x}\mathbf{B}$, the induction is simple. Now suppose **A** is $\Diamond\mathbf{B}$. If $\mathfrak{A}\models_k\Diamond\mathbf{B}[v]$, it is easy to see that $\mathfrak{B}\models_k\Diamond\mathbf{B}[v]$. Suppose on the other hand that $\mathfrak{B}\models_k\Diamond\mathbf{B}[v]$.

If $k \in \mathbf{L}-\mathbf{M}$, then since $k \in \mathbf{K}$, $k \in \mathrm{K}-\mathbf{N}$, and so $\mathfrak{A} \models_k \Diamond \mathbf{B}[v]$. Suppose that $k \in \mathbf{M}$, so that for some $l \in \mathbf{L}$ with $k\mathbf{S}l$, we have $\mathfrak{B} \models_l \mathbf{B}[v]$. If $l \in \mathbf{K}$, then $k\mathbf{R}l$ and by induction, $\mathfrak{A} \models_l \mathbf{B}[v]$, so that $\mathfrak{A} \models_k \Diamond \mathbf{B}[v]$. If l is p_γ, then since $k\mathbf{S}l$, then $k\mathbf{RO}$. By induction, since $\mathfrak{B} \models_{p_\gamma} \mathbf{B}[v]$, then $\mathfrak{B} \models_{\mathbf{p}} \mathbf{B}[v]$, and so again by induction, $\mathfrak{A} \models_{\mathbf{O}} \mathbf{B}[v]$. It follows that $\mathfrak{A} \models_k \Diamond \mathbf{B}[v]$.

Next, $\mathbf{P} \in \mathbf{M} \leftrightarrow p_\gamma \in \mathbf{M}$ for each $\gamma < \kappa$. So assume that $\mathbf{P} \notin \mathbf{M}$. If $\mathfrak{B} \models_{\mathbf{p}} \Diamond \mathbf{B}[v]$, then for some $l \in \mathbf{L}$ with $\mathbf{PS}l$, $\mathfrak{B} \models_l \mathbf{B}[v]$. If $l \in \mathbf{K}$, then by definition of $\mathbf{S}$, $p_\gamma \mathbf{S} l$ and so $\mathfrak{B} \models_{p_\gamma} \Diamond \mathbf{B}[v]$. If $l = p_\delta$, $\delta < \kappa$, we have $\mathbf{PS}p_\delta$, and so we have $\mathbf{ORO}$. Then $p_\gamma \mathbf{S} p_\delta$ and so $\mathfrak{B} \models_{p_\gamma} \Diamond \mathbf{B}[v]$. On the other hand, if $\mathfrak{B} \models_{p_\gamma} \Diamond \mathbf{B}[v]$, suppose $p_\gamma \in \mathbf{M}$, so that $\mathbf{P} \in \mathbf{M}$, and for some $l \in \mathrm{L}$ with $p_\gamma \mathbf{S} l$, $\mathfrak{B} \models_l \Diamond \mathbf{B}[v]$. If $l \in \mathbf{K}$, then $\mathbf{PS}l$, and so $\mathfrak{B} \models_{\mathbf{p}} \Diamond \mathbf{B}[v]$. If $l = p_\delta$, then $\mathbf{ORO}$, and then $\mathbf{PS}p_\delta$, so again $\mathfrak{B} \models_{\mathbf{p}} \Diamond \mathbf{B}[v]$, as desired. ■

A modal structure $\mathfrak{A} = \langle \mathscr{A}_k, \mathbf{K}, \ldots \rangle$ is said to be *universally infinite* if $|\mathscr{A}_k|$ is infinite for each $k \in \mathbf{K}$.

THEOREM 5.9. Let $\mathfrak{A}$ be a universally infinite modal structure with an infinite base such that for each $k \in \mathbf{K}$, $\mathrm{card}(\{k' \in \mathbf{K}: k'\mathbf{R}k \vee k' = k\}) \leq c(k)$. Let c be a constellation appropriate to $b(\mathfrak{A})$ such that $c^{\mathfrak{A}} \leq c$ and for all $k \in \mathbf{K}$, $\mathrm{card}(\mathrm{ML}) \leq c(k)$, and for all $k, k' \in \mathbf{K}$, if $k \neq k'$ and not $-k\mathbf{R}k'$ and $c(k) > c(k')$, then $|\mathscr{A}_k| - |\mathscr{A}_{k'}|$ is infinite. Then there exists a $\mathfrak{B}$ such that

(i) $\mathrm{id}: \mathfrak{A} \to \mathfrak{B}$.

(ii) $b(\mathfrak{A}) \prec b(\mathfrak{B})$ in the sense of classical first-order logic.

(iii) $c \leq c^{\mathfrak{B}} \upharpoonright \mathrm{dom}(c) \,\&\, \mathrm{card}(\mathrm{dom}(c^{\mathfrak{B}})) = \sup\{c(k): k \in \mathbf{K}\}$.

Proof. Construct the classical first-order structure $\mathfrak{A}^*$ described in Definition 5.5 and let T be the classical theory in $\mathrm{LS_T}$ whose nonlogical axioms consist of all formulas of $\mathrm{LS_T}(\mathfrak{A}')$ which are valid in $\mathfrak{A}' = \langle \mathfrak{A}^*, \mathbf{w}, \mathbf{i} \rangle_{w \in \mathrm{W}, i \in \mathrm{I}}$, where $\mathfrak{A}^* = \langle \mathrm{W}, \mathrm{I}, \ldots \rangle$ and the $\mathbf{w}$ for $w \in \mathrm{W}$ (resp. $\mathbf{i}$ for $i \in \mathrm{I}$) are individual constants of sort 1 (resp. sort 2) such that $\mathbf{w}_{\mathfrak{A}'} = w$ (resp. $\mathbf{i}_{\mathfrak{A}'} = i$). For each $k \in \mathbf{K}$, let $\{e^k_\delta : \delta < c(k)\}$ be a set of mutually distinct individual constants which do not occur in $\mathrm{LS_T}$ and are such that if $k \neq k'$, $\delta < c(k)$, and $\delta' < c(k')$, then c^k_δ is distinct from $c^{k'}_{\delta'}$. All the c^k_δ are to be constants of sort 2. Let T′ be the theory obtained from T by adding the following axioms:

(5.9) $e^k_\delta \neq e^{k'}_{\delta'}$, for $k \neq k'$ or $k = k'$ and $\delta \neq \delta', k, k' \in \mathbf{K}$, $\delta < c(k), \delta' < c(k')$,

(5.10) $\underline{k}'\mathrm{B}^* e^k_\delta$ for $k, k' \in \mathbf{K}$, $k\mathbf{R}k'$ or $k = k'$, and $\delta < c(k)$,

(5.11) $\neg \underline{k}'\mathrm{B}^* e^k_\delta$, for $k, k' \in \mathbf{K}$, not $-(k\mathrm{R}k'$ or $k \neq k')$, and $\delta < c(k)$ and $c(k) > c(k')$.

Let Σ be a finite subset of the Formulas (5.9)–(5.11). Since $\mathfrak{A}$ is universally infinite, it is clear that there is a selection of values in $\mathfrak{A}$ for the e^k_δ such that the formulas of the form (5.9) in Σ are satisfied, and moreover, that this selection can be made so as to satisfy the formulas in Σ of the forms (5.10) and (5.11), given the conditions relating c and $\mathfrak{A}$. Consequently, Σ is consistent, and hence so is T′. The usual Henkin constructions for single-sorted languages (cf. Shoenfield (1967) or Sacks (1972)) carry over easily to the two-sorted language of T′. Alternatively, one can first reduce the language of T′ to a single-sorted language (cf. Wang (1952) or (1970), Chapter XII) and then apply the usual completeness arguments (cf. Mal'cev (1971), Chapter 11, §2.2). By either method, we can construct a model of T″ in which each entity of each sort is named by a constant (of appropriate sort) in an extension T″ of T′ such that the cardinality of the set of constants of T″ is identical with the cardinality of the set of constants of T′.

Let $\mathfrak{C} = \langle \mathrm{W}, \mathrm{I}, \ldots \rangle$ be the model of T″ just described. We construct a modal structure $\mathfrak{B} = \langle \mathscr{B}_l, \mathbf{L}, \mathbf{S}, \mathbf{P}, \mathbf{M} \rangle$ as follows. Let $\mathbf{L} = \mathrm{W}$, $\mathbf{S} = \mathrm{R}_{*\mathfrak{C}}$, $\mathbf{P} = \mathrm{O}_{*\mathfrak{C}}$, and $\mathbf{M} = \mathrm{N}_{*\mathfrak{C}}$. For $l \in \mathbf{L} = \mathrm{W}$, let $|\mathscr{B}_l| = \{\mathbf{i} \in \mathrm{I} : l\mathrm{B}_{*\mathfrak{C}} i\}$. If $\mathbf{f}$ and $\mathbf{p}$ are n-ary function and predicate constants of ML, respectively, and $b_1, \ldots, b_n \in |\mathscr{B}_l|$, define $\mathbf{f}_{\mathscr{B}_l}(b_1, \ldots, b_n) = \mathbf{f}_{*\mathfrak{C}}(l, b_1, \ldots, b_n)$, and $\mathbf{p}_{\mathscr{B}_l}(b_1, \ldots, b_n)$ iff $\mathbf{p}_{*\mathfrak{C}}(l, b_1, \ldots, b_n)$. This specifies $\mathfrak{B}$. It is easy to see that $\mathfrak{B}^*$ is the *restriction* of $\mathfrak{C}$ to the language of T′ in the sense that the interpretations of the additional constants of T″ are dropped. Then for formulas **A** of the language of T′ and assignments v in $\mathfrak{B}^*$, it is easy to see that

$$\mathfrak{B}^* \models \mathbf{A}[v] \quad \text{iff} \quad \mathfrak{C} \models \mathbf{A}[v]$$

(cf. Shoenfield (1967), p. 43). Now for each $k \in \mathbf{K}$ and $a \in |\mathscr{A}_k|$, both $\underline{k}_\mathfrak{C}$ and $\underline{a}_\mathfrak{C}$ are elements of the appropriate domains of $\mathfrak{C}$. Then, if necessary, we can replace $\underline{k}_\mathfrak{C}$ by k and $\underline{a}_\mathfrak{C}$ by a, etc., so that we may regard $\mathfrak{A}^*$ as a substructure of $\mathfrak{C}$ and hence also of $\mathfrak{B}^*$. Then it follows that $\mathfrak{A}$ is a substructure of $\mathfrak{B}$. Now let **A** be any formula of ML, let $k \in \mathbf{K}$, and let v be an assignment in $\mathfrak{A}$. Suppose that $\mathfrak{A} \models_k \mathbf{A}[v]$. Then by Theorem 5.6, $\mathfrak{A}^* \models \mathbf{A}^*[v^*_k]$. Let **B** be the sentence which results from $\mathbf{A}^*$ by replacing each free variable $\mathbf{v}$ in $\mathbf{A}^*$ by the constant in the language of T which names $v^*_k(\mathbf{v})$. Then **B** is an axiom of T and so holds in T″. Hence **B** is valid in $\mathfrak{C}$ and hence also in $\mathfrak{B}^*$ by the remarks above. Then clearly $\mathfrak{B}^* \models \mathbf{A}^*[v^*_k]$ and so again by Theorem 5.6, $\mathfrak{B} \models_k \mathbf{A}[v]$. Since this holds in particular for formulas of the form $\neg \mathbf{A}$, it follows that $\mathfrak{A} \prec \mathfrak{B}$. Similarly, $b(\mathfrak{A}) \prec b(\mathfrak{B})$. Part (iii) follows easily from the details of the construction of $\mathfrak{C}$. ■

§6. ULTRAPRODUCTS

As in classical model theory, the construction of ultraproducts will be a very important tool in our development of modal model theory. Given an arbitrary non-empty index set I and a sequence $\langle \mathfrak{A}_i : i \in I \rangle$ of modal structures for the language ML, we define the *cartesian product* $X_{i \in I} \mathfrak{A}_i$ to be the following modal structure $\langle \mathscr{B}_l, \mathbf{L}, \mathbf{S}, \mathbf{P}, \mathbf{M} \rangle$ for ML. For each i, let $\mathfrak{A}_i = \langle \mathscr{A}^i_k, \mathbf{K}^i, \mathbf{R}^i, \mathbf{O}^i, \mathbf{N}^i \rangle$. Then set $\mathbf{L} = X_{i \in I} \mathbf{K}^i$ the ordinary set-theoretic cartesian product, and for $l, l' \in \mathbf{L}$, define $l\mathbf{S}l'$ iff $\forall i \in I [l(i) \mathbf{R}^i l'(i)]$. We set $\mathbf{P} = \langle \mathbf{O}^i : i \in I \rangle$ and let $\mathbf{M} = \{l \in \mathbf{L} : \forall i \in I [\mathbf{N}^i(l(i))]\}$. Finally, for $l \in \mathbf{L}$, let

$$\mathscr{B}_l = X_{i \in I} \mathscr{A}^i_l(i),$$

the classical cartesian or direct product of the structures $\mathscr{A}^i_{l(i)}$ (cf. Kopperman (1972), p. 75 or Bell and Slomsen (1971), p. 87ff). It is clear that $X_{i \in I} \mathfrak{A}_i$ is again a modal structure. Moreover, it is easy to verify that each of the formulas (of LB) N*(O*), CD1–CD14, and Norm is equivalent to a Horn formula. From this and Horn (1951) (cf. Shoenfield, pp. 94–95, problem 7), it follows that if each $\mathfrak{A}_i$ is an S-structure, then $X_{i \in I} \mathfrak{A}_i$ is again an S-structure. However, our main concern is not with cartesian products, but rather with ultraproducts. Given a non-empty set I, a *filter* F on I is a non-empty collection of subsets of I which has the two properties that $A, B \in F$ imply $A \cap B \in F$ and $A \in F \,\&\, A \subseteq B \subseteq I$ implies $B \in F$. The filter F is *proper* if $\emptyset \notin F$ and F is an *ultrafilter* if F is a maximal proper filter. This latter is equivalent to each of the two assertions that (a), for any $A \subseteq I$, either $A \in F$ or $I - A \in F$, and (b), for any $A, B \subseteq I$, if $A \cup B \in F$, then either $A \in \mathrm{F}$ or $B \in F$.

DEFINITION 6.1. Given an arbitrary non-empty index set I, a filter F on I, and a sequence $\langle \mathfrak{A}_i : i \in I \rangle$ of modal structures for the language ML, we define the *reduced direct product* $\Pi_{i \in I} \mathfrak{A}_i / F$ to be a structure $\mathfrak{B} = \langle \mathscr{B}_l, \mathbf{L}, \mathbf{S}, \mathbf{P}, \mathbf{M} \rangle$ for ML defined as follows. For each i let $\mathfrak{A}_i = \langle \mathscr{A}^i_k, \mathbf{K}^i, \mathbf{R}^i, \mathbf{O}^i, \mathbf{N}^i \rangle$. The base of $\mathfrak{B}$, $b(\mathfrak{B})$, is defined to be the classical reduced product

$$\langle \mathbf{L}, \mathbf{S}, \mathbf{P}, \mathbf{M} \rangle = \Pi_{i \in I} \langle \mathbf{K}^i, \mathbf{R}^i, \mathbf{O}^i, \mathbf{N}^i \rangle / F.$$

(Cf. Frayne *et al.* (1962) or Kopperman (1972), p. 75.) Recall that this means that the elements of $\mathbf{K}$ are equivalence classes $f/\sim_F$ where $f \in X_{i\in I}\mathbf{K}^i$ and for $f, g \in \mathbf{K}^i$, $f \sim_F g$ iff $\{i \in I : f(i) = g(i)\} \in F$. Also we have:

$$(f/\sim_F)\mathbf{S}(g/\sim_F) \quad \text{iff} \quad \{i \in I : f(i)\mathbf{R}^i g(i)\} \in F,$$

$$\mathbf{O} = \langle \mathbf{O}^i : i \in I \rangle / \sim_F,$$

$$f/\sim_F \in \mathbf{M} \quad \text{iff} \quad \{i \in I : f(i) \in \mathbf{N}^i\} \in F.$$

Now we must define $\mathscr{B}_l$ for $l \in \mathbf{L}$. Let $l = f/\sim_F$ and set

$$\bar{B}_l = \cup \{X_{i\in I} |\mathscr{A}^i_{g(i)}| : g \sim_F f\}.$$

Then for $h \in \bar{B}_l$, $h/\sim_F = \{h' \in \bar{B}_l : \{i \in I : h'(i) = h(i)\} \in F\}$, and we define

$$|\mathscr{B}_l| = \{h/\sim_F : h \in \underset{i\in I}{\times}\, \mathrm{U}(\mathfrak{A}_i) \,\&\, \exists \bar{h} \in \bar{B}_l [\bar{h} \sim_F h]\}.$$

If $\mathbf{p}$ and $\mathbf{f}$ are n-ary predicate and function symbols of ML, respectively, we define

$$\mathbf{p}_{\mathscr{B}_l}(h_1/\sim_F, \ldots, h_n/\sim_F \quad \text{iff} \quad \{i \in I : \mathbf{p}_{\mathscr{A}_{f(i)}}(h_1(i), \ldots, h_n(i))\} \in F,$$

and

$$\mathbf{f}_{\mathscr{B}_l}(h_1/\sim_F, \ldots, h_n/\sim_F) = \langle \mathbf{f}_{\mathscr{A}_{f(i)}}(h_1(i), \ldots, h_n(i)) : i \in I \rangle / \sim_F,$$

where in the latter expression, if not all of $h_1(i), \ldots, h_n(i)$ belong to $|\mathscr{A}_{f(i)}|$, then $\mathbf{f}_{\mathscr{A}_{f(i)}}(h_1(i), \ldots, h_n(i))$ is to be an arbitrary element of $|\mathscr{A}_{f(i)}|$. These interpretations are shown to be well-defined by arguments similar to those in the classical case. Note that one must not only show that these definitions are not only independent of the choice of representatives h_i of the classes $h_i/\sim_F$, but also of the choice of representative f of the class $l = f/\sim_F$.

This definition of ultraproduct is somewhat more complicated than a related definition given by Gabbay (1972b) for intuitionistic structures. This additional complication is necessary to guarantee that 'domains increase' in the ultraproduct and that the structures $\mathscr{B}_l$ are well-defined (cf. Bowen, 1973). We verify this now. Suppose that $(f/\sim_F)\mathbf{S}(g/\sim_F)$, $l = f/\sim_F$, and $h/\sim_F \in |\mathscr{B}_l|$. Then $h \in X_{i\in I}\mathrm{U}(\mathfrak{A}_i)$ and for some $\bar{h} \in \bar{B}_l$, $\bar{h} \sim_F h$. Consequently, for some $\bar{f} \sim_F f$, $\bar{h} \in X_{i\in I} |\mathscr{A}^i_{\bar{f}(i)}|$. Let $Z_0 = \{i \in I : f(i)\mathbf{R}^i g(i)\}$; then $Z_0 \in F$. Let $Z_1 = \{i \in I : \bar{f}(i) = f(i)\}$ so that $Z_1 \in F$. Define $\bar{g}$ by:

$$\bar{g}(i) = \begin{cases} g(i) & \text{if} \quad i \in Z_0 \cap Z_1, \\ f(i) & \text{otherwise.} \end{cases}$$

Let $z \in |\mathscr{A}^i_{\bar{f}(i)}|$. If $i \notin Z_0 \cap Z_1$, then $\bar{f}(i) = \bar{g}(i)$ and $z \in |\mathscr{A}^i_{\bar{g}(i)}|$. If $i \in Z_0 \cap Z_1$, then $\bar{f}(i) = f(i)\mathbf{R}^i g(i) = \bar{g}(i)$ so $z \in |\mathscr{A}^i_{\bar{g}(i)}|$. Thus for all i,

$$|\mathscr{A}^i_{f(i)}| \subseteq |\mathscr{A}^i_{\bar{g}(i)}|,$$

and so

$$\bar{h} \in X_{i\in I} |\mathscr{A}^i_{\bar{g}(i)}|.$$

But $\{i \in I: \bar{g}(i) = g(i)\} \supseteq Z_0 \cap Z_1 \in F$, so $\bar{g} \sim_F g$ and so $\bar{h} \in \bar{B}_l$, where $l' = g/\sim_F$. Hence $h/\sim_F \in |\mathscr{B}_{l'}|$ and so $|\mathscr{B}_l| \subseteq |\mathscr{B}_{l'}|$.

To see the necessity for this more complicated definition of $\mathscr{B}_l$, suppose that $l = f/\sim_F = g/\sim_F$, that $f(i_0) \neq g(i_0)$, and that $a_0 \in |\mathscr{A}^{i_0}_{f(i_0)}| - |\mathscr{A}^{i_0}_{g(i_0)}|$. Let $h \in X_{i\in I} |\mathscr{A}^i_{f(i)}|$ be such that $h(i_0) = a_0$. Then $h \notin X_{i\in I} |\mathscr{A}^i_{g(i)}|$ and so if we were to define $\mathscr{B}_l$ by $\mathscr{B}_l = \Pi_{i\in I} \mathscr{A}^i_{f(i)}/F$, $\mathscr{B}_l$ would not be well-defined with respect to l.

If for all $i \in I$, $\mathfrak{A}_i = \mathfrak{A}$, we write $\mathfrak{A}^I_F$ for the reduced product. As usual, if F is an ultrafilter, we call $\Pi_{i\in I}\mathfrak{A}_i/F$ an *ultraproduct* and $\mathfrak{A}^I_F$ an *ultrapower*. For any filter F, we define the *canonical embedding* g of $\mathfrak{A}$ into $\mathfrak{A}^I_F$ as follows. Let $\mathfrak{A} = \langle \mathscr{A}_k, \mathbf{K}, \mathbf{R}, \mathbf{O}, \mathbf{N} \rangle$. Then for $k \in \mathbf{K}$, set $g(k) = \langle k: i \in I \rangle / \sim_F$ and for $a \in |\mathscr{A}_k|$, let $g(a) = h_k(a)$, where h_k is the classical canonical embedding $h_k: \mathscr{A}_k \to (\mathscr{A}_k)^I_F$, noting that $X_{i\in I}|\mathscr{A}_k| \subseteq \bar{B}_l$, where $l = g(k)$. It is easy to verify that g is a monomorphism.

If v is an assignment in $X_{i\in I}\mathfrak{A}_i$, then for each variable $\mathbf{x}$, there is an $f \in X_{i\in I}\mathbf{K}^i$ such that $v(\mathbf{x})$ belongs to the universe of the structure corresponding to f; we define $\bar{v}$ by setting $\bar{v}(\mathbf{x}) = f$. Then we define v/F by $(v/F)(\mathbf{x}) = v(\mathbf{x})/\sim_F \in |\mathscr{B}_l|$, where $l = \bar{v}(\mathbf{x})/\sim_F$ and $\mathscr{B}_l$ is as in Definition 2.1. Then v/F is an assignment in $\Pi_{i\in I}\mathscr{A}_i/F$ and every assignment in $\Pi_{i\in I}\mathscr{A}_i/F$ can be obtained in this manner. Finally, we define $v|i$ by $(v|i)(\mathbf{x}) = v(\mathbf{x})(i)$.

LEMMA 6.2. Let Γ be a set of sentences of LB, let I be a non-empty index set, let F be an ultrafilter on I, and let $\langle \mathfrak{A}_i : i \in I \rangle$ be a sequence of modal structures for the language ML such that $\{i \in I: \mathfrak{A}_i \in \mathrm{St}(\mathrm{ML}, \Gamma)\} \in F$. Then $\Pi_{i\in I}\mathfrak{A}_i/F$ belongs to $\mathrm{St}(\mathrm{ML}, \Gamma)$.

Proof. Let $C = \{i \in I: \mathfrak{A}_i \in \mathrm{St}(\mathrm{ML}, \Gamma)\}$ and let $\mathfrak{B} = \Pi_{i\in I}\mathfrak{A}_i/F$. Let $\mathbf{A} \in \Gamma$. By Łoš's Theorem (cf. Bell and Slomsen (1971), p. 90 or Kopperman (1972), p. 77) $b(\mathfrak{B}) \models \mathbf{A}$ iff $\mathrm{D} = \{\in I: b(\mathfrak{A}_i) \models \mathbf{A}\} \in F$. Now $i \in C$ implies that $b(\mathfrak{A}_i) \models \mathbf{A}$, so that $C \subseteq D$. Since $C \in F$, then $D \in F$, and so $b(\mathfrak{B}) \models \mathbf{A}$. Since $\mathbf{A} \in \Gamma$ was arbitrary, it follows that $\mathfrak{B} \in \mathrm{St}(\mathrm{ML}, \Gamma)$. ■

COROLLARY 6.3. If S is one of the modal systems explicitly considered in § 2, the ultraproduct of S-structures is again an S-structure.

Note that the property of being a weak tree can be expressed by the formula

$$\forall w w_1 w_2 \blacksquare\ w_1 R^* w \,\&\, w_2 R^* w \rightarrow w_1 R^* w_2 \vee w_2 R^* w_1 .$$

Thus the ultraproduct of weak trees is again a weak tree.

The following theorem extends Łoš's Theorem to the present context.

THEOREM 6.4.* Let F be an ultrafilter on the non-empty index set I and let $\langle \mathfrak{A}_i : i \in I \rangle$ be a sequence of modal structures for ML. Let $\mathbf{A}$ be a formula of ML and let v be an assignment in $X_{i\in I}\mathfrak{A}_i$. Then for each $f \in X_{i\in I}\mathbf{K}^i$, if $l = f/\sim_F$,

$$\Pi_{i\in I}\mathfrak{A}_i/F \models_l \mathbf{A}[v/F] \quad \text{iff} \quad \{i \in I : \mathfrak{A}_i \models_{f(i)} \mathbf{A}[v|i]\} \in F .$$

Proof. The proof, as in the classical case, is by induction on the structure of $\mathbf{A}$ (cf. Bell and Slomsen, 1971). We will consider only the case where $\mathbf{A}$ is $\Diamond\mathbf{B}$. Let $\Pi_{i\in I}\mathfrak{A}_i/F = \langle \mathscr{B}_l, \mathbf{L}, \mathbf{S}, \mathbf{P}, \mathbf{M} \rangle = \mathfrak{B}$ and let $C = \{i \in I : \mathfrak{A}_i \models_{f(i)} \Diamond\mathbf{B}[v|i]\}$. Note that $l \in \mathbf{M}$ iff $D = \{i \in I : l(i) \in \mathbf{N}^i\} \in F$, and that $D \subseteq C$. Suppose $\mathfrak{B} \models_l \Diamond B[v/F]$. If $l \in \mathbf{M}$, then as we just noted, $C \in F$ as desired. If $l \notin M$, there is an $l' = f'/\sim_F$ such that $l\mathbf{S}l'$ and $\mathfrak{B} \models_{l'} \mathbf{B}[v/F]$. Let $\mathrm{E} = \{i \in I : f(i)\mathbf{R}^i f'(i)\}$, so that $E \in F$. By induction, if $\mathrm{G} = \{i \in I : \mathfrak{A}_i \models_{f'(i)} \mathbf{B}[v|i]\}$, then $G \in F$. Thus $E \cap G \in F$. But clearly, $E \cap G \subseteq C$, so $C \in F$. For the converse, assume that $C \in F$. Now $C = D \cup C - D$, and since F is an ultrafilter, it follows that either $D \in F$ or $C - D \in F$. As noted above, if $D \in F$, then $l \in \mathbf{M}$ and so $\mathfrak{B} \models_l \Diamond\mathbf{B}[v/F]$. Suppose $C - D \in F$, and let $i \in C - D$. Then $l(i) \notin \mathbf{N}^i$ and so there is a $k_i \in \mathbf{K}^i$ such that $f(i)\mathbf{R}^i k_i$ and $\mathfrak{A}_i \models_{k_i} \mathbf{B}[v \mid i]$. Define f' on I by:

$$f'(i) = \begin{cases} k_i & \text{if } \ i \in C - D \\ f(i) & \text{otherwise.} \end{cases}$$

Then if E is as above, $C - D \subseteq E$, so $E \in F$, and hence if $l' = f'/\sim_F$ then $l\mathbf{S}l'$. Let $H = \{i \in I : \mathfrak{A}_i \models_{f'(i)} \mathbf{B}[v \mid i]\}$. Then $C - D \subseteq H$ so that $H \in F$ and hence by induction, $\mathfrak{B} \models_{l'} \mathbf{B}[v/F]$. It now follows that $\mathfrak{B} \models_l \Diamond\mathbf{B}[v/F]$. ∎

THEOREM 6.5. The canonical embedding g: $\mathfrak{A} \rightarrow \mathfrak{A}_F^I$ is an elementary embedding whenever F is an ultrafilter.

Proof. Write $g^*(z) = \langle z : i \in I \rangle$ so that $g(z) = g^*(z)/\sim_F$. Then for any

* Note that all truth-valuations on atomic sentences over $\Pi_{i\in I}\mathfrak{A}_i/F$ satisfy the hypothesis of Proposition 4 of the Goldblatt (1975), so that the conflict between §3 of that paper and the present work is only apparent.

k, $\mathbf{A}$, and v we have

$$\mathfrak{A}^I_F \models_{g(k)} \mathbf{A}[g \circ v] \quad \text{iff} \quad \{i \in I : \mathfrak{A} \models_{g^*(k)(i)} \mathbf{A}[(g^* \circ v)|i]\} \in F$$
$$\text{iff} \quad \{i \in I : \mathfrak{A} \models_k \mathbf{A}[v]\} \in F$$
$$\text{iff} \quad \mathfrak{A} \models_k \mathbf{A}[v]. \blacksquare$$

DEFINITION 6.6. Let $\mathfrak{A}$ be a modal structure for ML. A *formula bundle over* $\mathfrak{A}$ is a finite function h such that:

(1) $\mathbf{O} \in \text{dom}(h) \subseteq \mathbf{K}$.

(2) for each $k \in \text{dom}(h)$, $h(k)$ is an ordered pair $\langle h_1(k), h_2(k) \rangle$ such that $h_1(k)$ is a finite subset of $|\mathscr{A}_k|$ and $h_2(k)$ is a finite set of closed formulas the form

$$\mathbf{A}_{\mathbf{x}_1, \ldots, \mathbf{x}_n}[\bar{a}_1, \ldots, \bar{a}_n],$$

where A is a formula of ML, $a_1, \ldots, a_n \in h_1(k)$, and for $i = 1, \ldots, n$, $\bar{a}_i$ is the canonical name for a_i in $L(\mathscr{A}_k)$ in the sense of Shoenfield (1967), p. 18, and if $v(\mathbf{x}_i) = a_i$ for $i = 1, \ldots, n$, then $\mathfrak{A} \models_k \mathbf{A}[v]$, and for each $a \in h_1(k)$, $\bar{a}$ occurs in $h_2(k)$.

For any finite set of formulas Z, let $\wedge Z$ be the conjunction of the formulas in Z.

DEFINITION 6.7. Let h be a formula bundle over the modal structure $\mathfrak{A}$ for ML. We define a series of formulas as follows: $\Psi_{h,k}$ is $\wedge h_2(k)$ and $\Psi'_{h,k}$ is obtained by replacing the canonical names $\bar{a}$ appearing in $\Psi_{h,k}$ by variables not occurring in $\Psi_{h,k}$, distinct names being replaced by distinct variables. Then $\theta_{h,k}$ is the existential closure of $\Psi'_{h,k}$. If $k \in \mathbf{K} - \mathbf{N}$, then $\theta'_{h,k}$ is $\theta_{h,k} \wedge \Diamond \exists \mathbf{x}[\mathbf{x} \neq \mathbf{x}]$. Otherwise, $\theta'_{h,k}$ is

$$\theta_{h,k} \wedge \Box \forall \mathbf{x}[\mathbf{x} = \mathbf{x}] \wedge \theta^{\#}_{h,k},$$

where $\theta^{\#}_{h,k}$ is

$$\wedge \{\Diamond \theta_{h,k'} : k' \in \text{dom}(h) \,\&\, k\mathbf{R}k'\}$$
$$\wedge \wedge \{\Diamond\Diamond \theta_{h,k'} : k' \in \text{dom}(h) \,\&\, k\mathbf{R}^2 k'\}$$
$$\wedge \wedge \{\Diamond^s \theta_{h,k'} : k' \in \text{dom}(h) \,\&\, k\mathbf{R}^s k'\},$$

where $s \leq \text{card}(\text{dom}(h))$ and $k\mathbf{R}^s k'$ means there are $k_1, \ldots, k_{s-1} \in \mathbf{K}$ such that

$$k\mathbf{R}k_1 \,\&\, k_1\mathbf{R}k_2 \,\&\, \cdots \,\&\, k_i\mathbf{R}k_{i+1} \,\&\, \cdots \,\&\, k_{s-1}\mathbf{R}k'.$$

Finally, θ_h is $\theta'_{h,\mathbf{O}}$ and θ^*_h is $\wedge h_2(\mathbf{O}) \wedge \Diamond \exists \mathbf{x}[\mathbf{x} \neq \mathbf{x}]$ if $\mathbf{O} \in \mathbf{K} - \mathbf{N}$ and otherwise, θ_h^* is

$$\wedge h_2(\mathbf{O}) \wedge \Box\forall\mathbf{x}[\mathbf{x}=\mathbf{x}] \wedge \theta^{\#}_{h,\mathbf{O}}.$$

If Γ is a collection of formulas of ML and h is a formula bundle over $\mathfrak{A}$, we say that h is a *Γ-formula bundle over* $\mathfrak{A}$ if for each $k\in\mathrm{dom}(h)$ and each formula $\mathbf{A}_{\mathbf{x}_1\ldots}[\bar{a}_1\ldots]$ in $h_2(k)$, the formula $\mathbf{A}$ belongs to Γ. A formula is *basic* if it is atomic or the negation of an atomic formula, and we say that h is *basic* if h is a Γ-formula bundle where Γ is the set of basic formulas of ML.

DEFINITION 6.8. If Γ is a set of formulas of ML and if $\mathfrak{A}$ is a modal structure for ML,

$$\mathrm{MD}^{\Gamma}(\mathfrak{A}) = \{\theta_h^* : h \text{ is a } \Gamma\text{-formula bundle over } \mathfrak{A}\}, \text{ and}$$

$$\widetilde{\mathrm{MD}}^{\Gamma}(\mathfrak{A}) = \{\theta_h : h \text{ is a } \Gamma\text{-formula bundle over } \mathfrak{A}\}.$$

If Γ is the set of all basic formulas of ML, we write $\mathrm{MD}(\mathfrak{A})$ for $\mathrm{MD}^{\Gamma}(\mathfrak{A})$ and call this set the *modal diagram* of $\mathfrak{A}$.

If h is a formula bundle over $\mathfrak{A}$ and $k\in\mathrm{dom}(k)$ it is easy to see that $\mathfrak{A}\models_k\theta'_{h,k}$. From this it follows that $\mathfrak{A}\models\theta_h^*$, and hence $\mathfrak{A}$ is a model of $\widetilde{\mathrm{MD}}^{\Gamma}(\mathfrak{A})$ and $\mathrm{MD}^{\Gamma}(\mathfrak{A})$ for any Γ.

DEFINITION 6.9. Let $\mathfrak{A}$ and $\mathfrak{B}$ be modal structures for ML and let Γ be a set of formulas of ML. A protomorphism m from $\mathfrak{A}$ to $\mathfrak{B}$ is a *Γ-morphism* (written $m:\mathfrak{A}\xrightarrow{\Gamma}\mathfrak{B}$) if for all $\mathbf{A}\in\Gamma$, all $k\in\mathbf{K}$, and all assignments v in $\mathfrak{A}$, $\mathfrak{A}\models_k\mathbf{A}[v]$ implies $\mathfrak{B}\models_{m(k)}\mathbf{A}[m\circ v]$.

Let $m:\mathfrak{A}\xrightarrow{\Gamma}\mathfrak{B}$. It is easy to check that m is a monomorphism iff Γ includes all basic formulas of ML, and that m is an elementary embedding iff Γ consists of all formulas of ML.

A set Γ of formulas is *regular* if it includes all formulas of the forms $\mathbf{x}=\mathbf{y}$ and $\neg\mathbf{x}=\mathbf{y}$, and contains $\mathbf{A}_{\mathbf{x}_1\ldots\mathbf{x}_n}[\mathbf{y}_1\ldots\mathbf{y}_n]$ whenever it contains $\mathbf{A}$.

If $\mathfrak{B}=\langle\mathscr{B}_l,\mathbf{L},\mathbf{S},\mathbf{P},\mathbf{M}\rangle$ is a modal structure for ML, $(\mathfrak{B},\underline{b})_{b\in|\mathscr{B}_{\mathbf{p}}|}=\mathfrak{B}'=\langle\mathscr{B}'_l,\mathbf{L},\mathbf{S},\mathbf{P},\mathbf{M}\rangle$ is that modal structure for ML′, where ML′ is ML together with the new individual constants $\underline{b}$ for each $b\in|\mathscr{B}_{\mathbf{p}}|$, obtained by expanding each $\mathscr{B}_l$ to a classical structure $\mathscr{B}'_l$ for L′ with the rule $\underline{b}_{\mathscr{B}'_l}=b$; i.e., $\mathscr{B}'_l$ is $(\mathscr{B}_{l'},\underline{b})_{b\in|\mathscr{B}_p|}$ (cf. Sacks, 1972).

DEFINITION 6.10. Let $\mathfrak{A}=\langle\mathscr{A}_k,\mathbf{K},\mathbf{R},\mathbf{O},\mathbf{N}\rangle$ and $\mathfrak{B}=\langle\mathscr{B}_l,\mathbf{L},\mathbf{S},\mathbf{P},\mathbf{M}\rangle$ be modal structures for ML. A *bundle from* $\mathfrak{A}$ *to* $\mathfrak{B}$ is a pair $\langle f,g\rangle$ of finite functions such that

(1) $\mathbf{O} \in \mathrm{dom}(f) \subseteq \mathbf{K}$, $f(\mathbf{O}) = \mathbf{P}$, if $k, k' \in \mathrm{dom}(f)$ and $k\mathbf{R}k'$, then $f(k)\,\mathbf{R}f(k')$, and for $k \in \mathrm{dom}(f)$, $k \in \mathbf{N}$ iff $f(k) \in \mathbf{M}$.

(2) $\mathrm{dom}(g) = \mathrm{dom}(f)$ and for each $k \in \mathrm{dom}(g)$, $g(k)$ is a finite function from a subset of $|\mathscr{A}_k|$ to $|\mathscr{B}_{f(k)}|$ such that if a, $a' \in \mathrm{dom}(g(k))$ and $a \equiv_k a'$, then $g(k)(a) \equiv_{f(k)} g(k)(a')$.

The next theorem appears to be the only workable replacement for the classical Diagram Lemma in the present setting.

THEOREM 6.11. Let $\mathfrak{A} = \langle \mathscr{A}_k, \mathbf{K}, \mathbf{R}, \mathbf{O}, \mathbf{N} \rangle$ with $\mathbf{O} \in \mathbf{N}$ and $\mathfrak{B} = \langle \mathscr{B}_l, \mathbf{L}, \mathbf{S}, \mathbf{P}, \mathbf{M} \rangle$ be modal structures for ML such that $\mathfrak{A}$ is a weak tree and $|\mathscr{A}_{\mathbf{o}}| \subseteq |\mathscr{B}_{\mathbf{P}}|$. If $\mathfrak{B}$ is a model of $\mathrm{MD}^{\Gamma}(\mathfrak{A})$ where Γ is a regular set of formulas, then there exist a set I, an ultrafilter F on I, and a Γ-morphism $m : \mathfrak{A} \xrightarrow{\Gamma} \mathfrak{B}^I_F$.

Proof. Let I be the set of all bundles from $\mathfrak{A}$ to $\mathfrak{B}$. For any Γ-formula bundle h over $\mathfrak{A}$, define $J_h \subseteq I$ to consist of all $\langle f, g \rangle$ such that

(i) $\mathrm{dom}(h) \subseteq \mathrm{dom}(f) = \mathrm{dom}(g)$;

(ii) $k, k' \in \mathrm{dom}(h)$ implies $h_1(k) \subseteq \mathrm{dom}(g(k))$ and $k\mathbf{R}k'$ implies $f(k)\mathbf{S}f(k')$;

(iii) if $k \in \mathrm{dom}(h)$ and $\mathbf{A}_{\mathbf{x}_1 \ldots \mathbf{x}_n}[\bar{a}_1, \ldots, \bar{a}_n] \in h_2(k)$, then $\mathfrak{B} \models_{f(k)} \mathbf{A}[\mu]$ where $\mu(\mathbf{x}_i) = g(k)(a_i)$ for $i = 1, \ldots, n$.

First we claim that for any Γ-formula bundle h, $J_h \neq \emptyset$. Since $\theta_h \in \mathrm{MD}^{\Gamma}(\mathfrak{A})$, then $\mathfrak{B} \models_{\mathbf{P}} \theta_h$. Since $\mathbf{O} \in \mathbf{N}$, then θ_h is

$$\theta_{h,\mathbf{O}} \wedge \Box\forall\mathbf{x}[\mathbf{x} = \mathbf{x}] \wedge \theta^{\#}_{h,\mathbf{O}},$$

so $\mathfrak{B} \models_{\mathbf{P}} \Box\forall\mathbf{x}[\mathbf{x} = \mathbf{x}]$ and so $\mathbf{P} \in \mathbf{M}$. Then since $\mathfrak{B} \models_{\mathbf{P}} \theta^{\#}_{h,\mathbf{O}}$ for each $s \leq \mathrm{card}(\mathrm{dom}(h))$ and each $k \in \mathrm{dom}(h)$ with $\mathbf{O}\mathbf{R}^s k$, we have $\mathfrak{B} \models_{\mathbf{P}} \Diamond^s \theta_{h,k}$. Since $\mathbf{P} \in \mathbf{M}$, there must be an $l \in \mathbf{L}$ such that $\mathfrak{B} \models_l \theta_{h,k}$. Since $\mathfrak{A}$ is a weak tree, l can be chosen so that (ii) is satisfied. Thus define f on $\mathrm{dom}(h)$ so that $f(k)$ is such an l. Suppose $\theta_{h,k}$ is $\exists x_1 \ldots x_\alpha \Psi'_{h,k}$. Thus there must be an assignment μ in $\mathfrak{B}$ such that $\mu(x_i) \in |\mathscr{B}_{f(k)}|$ for $i = 1, \ldots, \alpha$ and $\mathfrak{B} \models_{f(k)} \Psi'_{h,k}[\mu]$. Then for each a whose name $\bar{a}$ occurs in a formula of $h_2(k)$, define $g(k)(a) = \mu(x_i)$ where x_i is the variable which replaced $\bar{a}$ in the construction of $\Psi'_{h,k}$ from $\Psi_{h,k}$. It follows immediately that $\langle f, g \rangle \in J_h$.

For $k \in \mathbf{K}$ and $W \in S_\omega(|\mathscr{A}_k|)$, set

$$J_{k,\mathrm{w}} = \{\langle f, g \rangle \in I : k \in \mathrm{dom}(f)\ \&\ W \subseteq \mathrm{dom}(g(k))\}.$$

Then set

$$N = \{J_h : h \text{ is a } \Gamma\text{-formula bundle over } \mathfrak{A}\} \cup \{J_{k,W} : k \in \mathbf{K} \,\&\, W \in S_\omega(|\mathscr{A}_k|)\}.$$

We claim that N has the finite intersection property. Let $h^1, \ldots, h^n$ be Γ-formula bundles over $\mathfrak{A}$, let $k_1, \ldots, k_m \in \mathbf{K}$ and let $W_i \in S_\omega(|\mathscr{A}_{k_i}|)$ for $i = 1, \ldots, m$. Set

$$\mathrm{dom}(h) = \bigcup_{i=1}^{n} \mathrm{dom}(h^i) \cup \{k_1, \ldots, k_m\}.$$

For $k \in \mathrm{dom}(h)$, let $V_k = \cup_{j=1}^{n} {}^*h_2^j(k)$ where ${}^*h_2^j(k)$ consists of all $a \in |\mathscr{A}_k|$ such that $\bar{a}$ occurs in some formula of $h_2^j(k)$. Then define

$$h_1(k) = \begin{cases} V_k \cup W_s & \text{if } k = k_s \text{ for some } s = 1, \ldots, m \\ V_k & \text{otherwise.} \end{cases}$$

Let $Y_k = \bigcup_{j=1}^{n} H_2^j(k)$, and define

$$h_2(k) = \begin{cases} Y_k \cup \{\bar{a} = \bar{a} : a \in W_s\} & \text{if } k = k_s, s = 1, \ldots, m \\ Y_k & \text{otherwise.} \end{cases}$$

Then h is a Γ-formula bundle over $\mathfrak{A}$ and we have

$$\emptyset \neq J_h \subseteq \bigcap_{i=1}^{n} J_{h_i} \cap \bigcap_{j=1}^{m} J_{k_j}, W_j.$$

Thus N has the finite intersection property and so can be extended to an ultrafilter F on I. Define $\bar{m} : \mathbf{K} \to \mathbf{L}^I$ by

$$\bar{m}(k)(\langle f, g \rangle) = \begin{cases} f(k) & \text{if } k \in \mathrm{dom}(f) \\ \mathbf{P} & \text{otherwise.} \end{cases}$$

Note that for all $\langle f, g \rangle \in I$, $\bar{m}(\mathbf{O})(\langle f, g \rangle) = \mathbf{P}$. Now define $m : \mathbf{K} \to \mathbf{L}_F^I$ by $m(k) = \bar{m}(k)/F$. Let $b\#$ be a fixed (arbitrarily chosen) element of $|\mathscr{B}_\mathbf{P}|$. Next for $k \in \mathbf{K}$, define

$$\bar{m} : |\mathscr{A}_k| \to \cup \{X_{i \in I} |\mathscr{B}_{q(i)}| : q \in \mathbf{L}^I \,\&\, q \in m(k)/F\}$$

by

$$\bar{m}(a)(\langle f, g \rangle) = \begin{cases} g(k)(a) & \text{if } k \in \mathrm{dom}(g) \,\&\, a \in \mathrm{dom}(g(k)) \\ b\# & \text{otherwise,} \end{cases}$$

and define m on $|\mathscr{A}_k|$ by $m(a) = \bar{m}(a)/F$.

Now suppose that $a, a' \in |\mathscr{A}_k|$ with $a \not\equiv_k a'$. Let h be any Γ-formula bundle with $k \in \mathrm{dom}(h)$, $a, a' \in h_1(k)$, and $\bar{a} \neq \bar{a}' \in h_2(k)$. Let

$$Z = \{\langle f, g\rangle \in I : k \in \mathrm{dom}(f) \,\&\, a, a' \in \mathrm{dom}(g(k)) \\ \&\, \mathfrak{B} \models_{f(k)} x \neq y[g(k)(a), g(k)(a')]\}.$$

Then $J_h \subseteq Z$ so that $Z \in F$. Since, if $l = m(k)(i)$,

$$Z \subseteq \{i \in I : \overline{m}(a)(i) \not\equiv_l \overline{m}(a')(i)\},$$

it follows that $m(a) \not\equiv_{m(k)} m(a')$. Clearly $m(\mathbf{O}) = \mathbf{P}/F$. Next suppose that $k\mathbf{R}k'$, and let

$$Y = \{\langle f, g\rangle \in \mathrm{I} : k, k' \in \mathrm{dom}(f)\}.$$

From the argument that N has the finite intersection property, it is clear that for some h, $J_h \subseteq Y$ so that $Y \in F$. But

$$Y \subseteq \{i \in I : \overline{m}(k)(i)\mathbf{S}\overline{m}(k')(i)\},$$

and hence $m(k)(\mathbf{S}/F)m(k')$. That $k \in \mathbf{K} - \mathbf{N}$ iff $m(k) \in \mathbf{K}_F^I - \mathbf{N}_F^I$ is argued similarly. Now let $\mathbf{f}$ be an n-ary function symbol of ML and let $a_1, \ldots, a_n \in |\mathscr{A}_k|$. Let $c = \mathbf{f}_{\mathscr{A}_k}(a_1, \ldots, a_n)$ and let h be a Γ-formula bundle such that $k \in \mathrm{dom}(h)$, $a_1, \ldots, a_n, c \in h_1(k)$, and $\bar{c} = \mathbf{f}\,\bar{a}_1 \ldots \bar{a}_n \in h_2(k)$. Suppose $\langle f, g\rangle \in J_h$. Then $\overline{m}(c)(\langle f, g\rangle) = g(k)(c)$. But

$$\mathfrak{B} \models_{f(k)} \mathbf{y} = \mathbf{f}x_1 \ldots x_n[g(k)(c), g(k)(a_1), \ldots, g(k)(a_n)],$$

and it follows that

$$g(k)(c) \equiv_l \mathbf{f}_{\mathscr{B}_{f(k)}}(g(k)(a_1), \ldots, g(k)(a_n)) \\ = \mathbf{f}_{\mathscr{B}_{\overline{m}(k)}}(\overline{m}(a_1)(\langle f, g\rangle), \ldots, \overline{m}(a_n)(\langle f, g\rangle)),$$

where $l = \overline{m}(k)(\langle f, g\rangle)$. It follows that

$$J_h \subseteq \{\langle f, g\rangle \in I : \overline{m}(c)(\langle f, g\rangle) \\ \equiv_l \mathbf{f}_{\mathscr{B}_{\overline{m}(k)}}(\overline{m}(a_1)(\langle f, g\rangle), \ldots \overline{m}(a_n)(\langle f, g\rangle))\}$$

and so $m(f_{\mathscr{A}_k}(a_1, \ldots, a_n)) \equiv_{m(k)} \mathbf{f}_{\mathscr{B}_{m(k)}}(m(a_1), \ldots, m(a_n))$. By a similar argument,

$$\mathbf{p}_{\mathscr{A}_k}(a_1, \ldots, a_n) \leftrightarrow \mathbf{p}_{\mathscr{B}_{m(k)}}(m(a_1), \ldots, m(a_n))$$

for any predicate symbol $\mathbf{p}$. More generally, let $\mathbf{A} \in \Gamma$ have $\mathbf{x}_1, \ldots, \mathbf{x}_n$ as free variables, let $k \in \mathbf{K}$, and let $a_1, \ldots, a_n \in |\mathscr{A}_k|$.

Let h be a Γ-formula bundle such that $k \in \mathrm{dom}(h)$, $a_1, \ldots, a_n \in h_1(k)$, and $\mathbf{A}[\bar{a}_1, \ldots, \bar{a}_n] \in h_2(k)$, assuming $\mathfrak{A} \models_k \mathbf{A}[a_1, \ldots, a_n]$. Then

$$\emptyset \neq J_h \subseteq \{\langle f, g\rangle : k \in \mathrm{dom}(f) \,\&\, a_1, \ldots, a_n \in \mathrm{dom}(g(k)) \\ \&\, \mathfrak{B} \models_{f(k)} \mathbf{A}[g(k)(a_1), \ldots, g(k)(a_n)]\},$$

and it follows by Theorem 6.4 that

$$\mathfrak{B}^I_F \models_{m(k)} \mathbf{A}[m(a_1), \ldots, m(a_n)].$$

Thus m is a Γ-embedding of $\mathfrak{A}$ in $\mathfrak{B}^I_F$. ■

We apparently cannot guarantee that the Γ-morphism $m: \mathfrak{A} \to \mathfrak{B}^I_F$ is strong; i.e., $1-1$ on the set $\mathbf{K}$ of worlds of $\mathfrak{A}$ (cf. Bowen, 1975, pp. 115–116). However, we will see that we can achieve this if we replace $\mathfrak{B}$ by a closely related structure.

DEFINITION 6.12. Let $\mathfrak{B} = \langle \mathscr{B}_l, \mathbf{L}, \mathbf{S}, \mathbf{P}, \mathbf{M} \rangle$ and $\mathfrak{D} = \langle \mathscr{D}_z, \mathbf{Z}, \mathbf{U}, \mathbf{X}, \mathbf{Y} \rangle$ be S-structures. We say that $\mathfrak{D}$ *is a trivial expansion of* $\mathfrak{B}$ provided the following hold:

(1°) $\mathbf{L} \subseteq \mathbf{Z}$, $\mathbf{U} \upharpoonright \mathbf{L}^2 = \mathbf{S}$, $\mathbf{P} = \mathbf{X}$;

(2°) for all $z \in \mathbf{Z} - \mathbf{L}$, there is an $l \in \mathbf{L}$ such that:

(a) $\mathscr{D}_z = \mathscr{B}_l$;

(b) $z\mathbf{U}z$ iff $l\mathbf{S}l$;

(c) for all $l' \in L$, $z\mathbf{U}l'$ iff $l\mathbf{S}l'$, and $l'\mathbf{U}z$ iff $l'\mathbf{S}l$;

(d) $z \in \mathbf{Y}$ iff $l \in \mathbf{M}$;

(3°) if $z, z' \in \mathbf{Z} - \mathbf{L}$ correspond to l, l' respectively as in (2°), then $z\mathbf{U}z'$ iff $l\mathbf{S}l'$.

We will say that z is a *copy* of l above if z and l satisfy condition (2°) above, and we will write $\mathrm{cp}_{\mathfrak{D}}(l)$ for the set of copies of l. We will say that $\mathfrak{D}$ is a *κ-trivial expansion* of $\mathfrak{B}$ if for each $l \in \mathbf{L}$, $\mathrm{card}(\mathrm{cp}_{\mathfrak{D}}(l)) \geq \kappa$. It is obvious that the identity map is an elementary embedding of $\mathfrak{B}$ in $\mathfrak{D}$. Also, κ-trivial expansions of $\mathfrak{B}$ clearly exist for any κ.

THEOREM 6.13. Let $\mathfrak{A}$, $\mathfrak{B}$, and Γ be as in Theorem 6.11. Then there exist a set I, an ultrafilter F on I, a trivial expansion $\mathfrak{D}$ of $\mathfrak{B}$, and a strong (faithful) Γ-morphism $m: \mathfrak{A} \to \mathfrak{D}^I_F$ ($m: \mathfrak{A} \to E(\mathfrak{B})^I_F$).

Proof. Let $\mathfrak{D}$ be a $\aleph_0$-trivial expansion of $\mathfrak{B}$. Let I be the set of all bundles from $\mathfrak{A}$ to $\mathfrak{D}$, and define J_h as in the proof of Theorem 6.11 except that $\mathfrak{B}$ is replaced by $\mathfrak{D}$ and the map f is required to be $1-1$. It is clear that in the argument that $J_h \neq \emptyset$ we can easily construct an f which is $1-1$ since $\mathrm{dom}(f)$ is finite and each $l \in \mathbf{L}$ has $\aleph_0$ copies in $\mathfrak{D}$. Now proceed as in the proof of Theorem 6.11. To see that the m constructed is strong, suppose that $k, k' \in \mathbf{K}$ and $k \neq k'$. Then $m(k) \neq m(k')$ iff $\{i \in I : \overline{m}(k)(i) \neq \overline{m}(k')(i)\} \in \mathrm{F}$. Let $\mathrm{Z} = \{\langle f, g \rangle \in \mathrm{I} : k, k' \in \mathrm{dom}(f)\}$. From the argument that N has the finite intersection property, we see that $Z \in F$. But if $i = \langle f, g \rangle \in Z$, then

$$\overline{m}(k)(i) = f(k) \neq f(k') = \overline{m}(k')(i),$$

and so $\{i \in I : \overline{m}(k)(i) \neq \overline{m}(k')(i)\} \in F$. Thus m is $1-1$. To see that m can be made faithful when $E(\mathfrak{B}) = \mathfrak{D}$, define J_n as above, but require f to be faithful instead of $1-1$. ■

Now we apply ultrapowers to obtain new proofs of the Upward Löwenheim–Skolem–Tarski Theorem. A map $m: \mathfrak{A} \to \mathfrak{B}$ is said to be *exact* if $m \upharpoonright b(\mathfrak{A})$ is an isomorphism of $b(\mathfrak{A})$ on $b(\mathfrak{B})$ in the usual classical sense. An S-structure $\mathfrak{A}$ has a *finite base* if $b(\mathfrak{A})$ is a finite structure; moreover, $\mathfrak{A}$ is said to be *universally infinite* if for each $k \in \mathbf{K}$, $|\mathscr{A}_k|$ is infinite.

THEOREM 6.14. (Upward Löwenheim–Skolem–Tarski, I). Let $\mathfrak{A} = \langle \mathscr{A}_k, \mathbf{K}, \mathbf{R}, \mathbf{O}, \mathbf{N} \rangle$ be a universally infinite S-modal structure for ML such that $b(\mathfrak{A})$ is finite. Let c be a cardinal constellation with domain $\mathbf{K}$ such that $c^{\mathfrak{A}} \leq c$ and c is appropriate for $\mathfrak{A}$. Then there exists an S-modal structure $\mathfrak{B}$ and an exact map $m: \mathfrak{A} \to \mathfrak{B}$ such that m is an elementary embedding and $c^{\mathfrak{B}} = c$.

Proof. Let $\gamma = \sup\{c(k) : k \in K\}^+$. We may assume that $\gamma \geq \aleph_0$ since otherwise the theorem is trivial. Let I be an index set of power γ and let F be a regular ultrafilter on I (cf. Bell and Slomsen (1971), p. 114). Let $\mathfrak{C} = \mathfrak{A}^I_F$. Then $b(\mathfrak{C}) = b(\mathfrak{A})^I_F$ and since $b(\mathfrak{A})$ was finite, it follows that $b(\mathfrak{C})$ is isomorphic to $b(\mathfrak{A})$ (cf. Bell and Slomsen (1971), p. 126). So write $\mathfrak{C} = \langle \mathscr{C}_k, \mathbf{K}, \mathbf{R}, \mathbf{O}, \mathbf{N} \rangle$, $\mathscr{C}_k = \langle C_k, \equiv_k, \ldots \rangle$. Let $d_k \in \mathbf{K}^I$ be the constant function $d_k(i) = k$ for all $i \in I$. Then by construction of our ultraproduct, it follows that $\text{card}(|\mathscr{A}_k|^I_F) \leq \text{card}(C_k)$. But since F is regular and $\text{card}|\mathscr{A}_k| = c^{\mathfrak{A}}(k) \geq \aleph_0$, it follows that $\text{card}(|\mathscr{A}_k|^I_F) = (c^{\mathfrak{A}}(k))^\gamma$ (cf. Bell and Slomsen (1971), p. 132). And since γ is a regular cardinal, it follows by the Corollary to Konig's Theorem (Jech (1971), p. 17) that $(c^{\mathfrak{A}}(k))^\gamma > \gamma$. Now let F be a function with domain $\mathbf{K}$ such that for each k, $m''|\mathscr{A}_k| \subseteq F(k) \subseteq C_k$ and $\text{card}(F(k)) = c(k)$, where m is the canonical embedding of $\mathfrak{A}$ in $\mathfrak{A}^I_F$. Then it is clear that $c \leq c^{\mathfrak{A}}$, c is appropriate for $b(\mathfrak{C}) = b(\mathfrak{A})$, and for each $k \in \mathbf{K}$, $\max\{\text{card}(F(k)), \text{card}(\text{ML})\} = c(k)$ and $\text{card}\{k' \in \mathbf{K} : k'\mathbf{R}k\} < c(k)$, the latter since $\mathbf{K}$ is finite and $c(k) \geq \aleph_0$. Then by Theorem 5.1 there exists an S-modal structure $\mathfrak{B} = \langle \mathscr{B}_k, \mathbf{K}, \mathbf{R}, \mathbf{O}, \mathbf{N} \rangle$, $\mathscr{B}_k = \langle |\mathscr{B}_k|, \equiv_k, \ldots \rangle$, such that $c^{\mathfrak{B}} = c$, for each $k \in K$, $m'' A_k \subseteq F(k) \subseteq |\mathscr{B}_k|$, and $\text{id}: \mathfrak{B} \xrightarrow{=} \mathfrak{C}$, where id is the identity map. Let v be an assignment in $\mathfrak{A}$, let $\mathbf{A}$ be a formula of ML, and let $k \in \mathbf{K}$. Then

$$\begin{aligned} \mathfrak{A} \models_k \mathbf{A}[v] \quad &\text{iff} \quad \mathfrak{C} \models_k \mathbf{A}[\mathrm{m} \circ v], \\ &\text{iff} \quad \mathfrak{B} \models_k \mathbf{A}|\mathrm{m} \circ v]. \end{aligned}$$

Thus $m: \mathfrak{A} \xrightarrow{=} \mathfrak{B}$, as desired. ■

THEOREM 6.15. (Upward Löwenheim–Skolem–Tarski, II). Let $\mathfrak{A} = \langle \mathscr{A}_k, \mathbf{K}, \mathbf{R}, \mathbf{O}, \mathbf{N} \rangle$ be a universally infinite S-modal structure for ML (no restriction on $b(\mathfrak{A})$). Let c be a cardinal constellation with domain $\mathbf{K}$ such that $c^{\mathfrak{A}} \leq c$, c is appropriate for $\mathfrak{A}$, and for each $k \in \mathbf{K}$, card $\bar{\mathbf{R}}''k \leq c(k)$. Then there exists an S-modal structure $\mathfrak{B}$ for ML and a monomorphism $m: \mathfrak{A} \to \mathfrak{B}$ such that m is elementary and one-one on $\mathbf{K}$ and for $k \in \mathbf{K}$, $c^{\mathfrak{B}}(m(k)) = c(k)$.

Proof. Let $\gamma = \sup\{c(k): k \in \mathbf{K}\}^+$. Then $\gamma \geq \aleph_0$. Let I be an index set of power γ and let F be a regular ultrafilter on I. Let $\mathfrak{C} = \mathfrak{A}^I_F = \langle \mathscr{C}_m, \mathbf{M}, \mathbf{T}, \mathbf{Q}, \mathbf{X} \rangle$. For $k \in \mathbf{K}$, let $d_k: I \to \mathbf{K}$ be the constant function $d_k(i) = k$, and let $k^* = d_k/F$. By construction of the ultraproduct, $\text{card}(|\mathscr{A}_k|^I_F) \leq \text{card}(|\mathscr{C}_{k^*}|)$. Then as in the proof of Theorem 6.14, it follows that $\gamma < \text{card}(|\mathscr{C}_{k^*}|)$. Let m be the canonical embedding of $\mathfrak{A}$ in $\mathfrak{C}$ (note that $m(k) = k^*$). Let $J = m''\mathbf{K}$ and let H be a function with domain J such that $m''|\mathscr{A}_k| = H(k^*)$. Define c on J by $c(k^*) = c(k)$. Suppose $k^*, l^* \in J$ and $k^*\mathbf{M}l^*$. Then $k\bar{\mathbf{R}}l$ and so there are $k_1, \ldots, k_p \in \mathbf{K}$ such that $k = k_1, l = k_p$ and for $i = 1, \ldots, p-1$, $k_i\mathbf{R}k_{i+1}$. But then $c(k_i) \leq c(k_{i+1})$ for $i = 1, \ldots, p-1$, and it follows by the transitivity of $\leq$ that $c(k) < c(l)$ and so $c(k^*) \leq c(l^*)$. Then the hypotheses of Theorem 5.2 are met and so there is an S-modal structure $\mathfrak{B} = \langle \mathscr{B}_l, \mathbf{L}, \mathbf{S}, \mathbf{P}, \mathbf{M} \rangle$ as described there. From the definition of J and H, it is clear that m is a monomorphism of $\mathfrak{A}$ into $\mathfrak{B}$ and from Theorem 5.2, we get $c^{\mathfrak{B}}(m(k)) = c(k)$ for $k \in \mathbf{K}$. Finally, let $k \in \mathbf{K}$, let v be an assignment in $\mathfrak{A}$, and let $\mathbf{A}$ be a formula of ML. Then by conclusion (4°) of Theorem 4.7, we have:

$$\begin{aligned} \mathfrak{A} \models_k \mathbf{A}[v] \quad &\text{iff} \quad \mathfrak{C} \models_{m(k)} \mathbf{A}[m \circ v] \\ &\text{iff} \quad \mathfrak{B} \models_{m(k)} \mathbf{A}[m \circ v]. \end{aligned}$$

Hence

$$m: \mathfrak{A} \mathrel{\overset{=}{\Rightarrow}} \mathfrak{B}. \quad \blacksquare$$

§7. ULTRAFILTER PAIRS AND ELEMENTARY EMBEDDINGS

An ultrafilter pair $\langle I, F\rangle$ consists of a non-empty set Γ and an ultrafilter F on I. The next two theorems extend results of Frayne *et al.* (1962).

THEOREM 7.1. Let $\mathfrak{A}$ and $\mathfrak{B}$ be modal structures for ML. Then $\mathfrak{A} \equiv \mathfrak{B}$ iff there exists an ultrafilter pair $\langle I, F\rangle$ such that $\mathfrak{A}$ is elementarily embeddable in $\mathfrak{B}^I_F$.

Proof. Sufficiency is obvious. For necessity, let Γ consist of all formulas of ML, and note now that a Γ-embedding is an elementary embedding. Since $\mathfrak{A} \equiv \mathfrak{B}$, then $(\mathfrak{B}, b)^b_{b\in|\mathscr{B}_p|}$ is a model of $\mathrm{MD}^\Gamma(\mathfrak{A})$ and the result follows by Theorem 6.11. ∎

THEOREM 7.2. Let $\mathfrak{A}$ and $\mathfrak{B}$ be modal structures such that $\mathfrak{A}$ is a weak tree. A (strong) faithful monomorphism m from $\mathfrak{B}$ to $\mathfrak{A}$ is an elementary embedding iff there exist an ultrafilter pair $\langle I, F\rangle$ and a (strong) elementary faithful embedding p of $\mathfrak{A}$ in $\mathfrak{B}^I_F$ such that Figure 2 is weakly commutative, where q is the canonical embedding of $\mathfrak{B}$ in $\mathfrak{B}^I_F$, in the sense that if $l \in \mathbf{L}$, then $g(l) = (\mathrm{p} \circ \mathrm{m})(l)$, and if $b \in |\mathscr{B}_l|$, then $g(b) \equiv_{g(l)} (\mathrm{p} \circ \mathrm{m})(b)$; we write $g \approx \mathrm{p} \circ \mathrm{m}$ to indicate this:

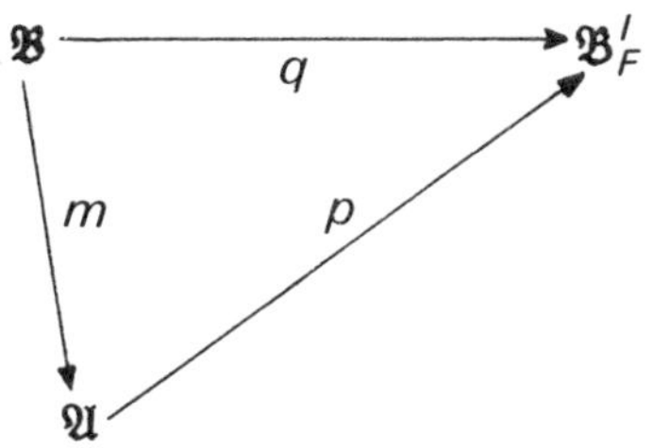

Fig. 2.

Proof. Sufficiency is obvious. For necessity, assume that $m: \mathfrak{B} \to \mathfrak{A}$ is elementary. We say that a bundle $\langle f, g\rangle$ from $\mathfrak{A}$ to $\mathfrak{B}$ is an m-bundle if for each $k \in \mathbf{K}$, if $k \in \mathrm{rng}(m)$, then $f(k) \in m^{-1}(k)$, and if $k \in \mathrm{dom}(f) \cap \mathrm{rng}(m)$ and $a \in |\mathscr{A}_k| \cap \mathrm{dom}(g) \cap \mathrm{rng}(m)$, then $g(\langle a\rangle) \in m^{-1}(a)$. Let I be the set of all strictly finite m-bundles over $\mathfrak{A}$ to $\mathfrak{B}$. Similarly, a formula bundle h over $\mathfrak{A}$ is an *m-formula bundle over* $\mathfrak{A}$ if $\mathrm{dom}(h) \subseteq \mathrm{rng}(m)$ and for each

$k \in \text{dom}(h)$, if $h(k) = \langle h_1(k), h_2(k) \rangle$, then $h_1(k) \subseteq |\mathscr{A}_k| \cap \text{rng}(m)$. Then define $J_h \subseteq I$ just as $J_h \subseteq I$ was defined in the proof of Theorem 6.11. Given an m-formula bundle h, to see that $J_h \neq \emptyset$, proceed as follows. f will be a finite function with domain $\text{dom}(h)$ such that if $k \in \text{dom}(f) \cap \text{rng}(m)$, $f(k) \in m^{-1}(k)$. Now let $k \in \text{dom}(f) - \text{rng}(m)$. Let $\theta_{h,k}$ be as constructed in the proof of Theorem 6.1. Then $\mathfrak{A} \models_{\mathbf{O}} \Diamond \theta_{h,k}$; since $\mathfrak{B} \equiv \mathfrak{A}$, then $\mathfrak{B} \models_{\mathbf{P}} \Diamond \theta_{h,k}$, so for some $l \in \mathbf{L}$, $\mathfrak{B} \models_l \theta_{h,k}$; let $f(k) = l$. Now $\theta_{h,k}$ is $\exists \mathbf{x}_0 \ldots \exists \mathbf{x}_q \Psi'_{h,k}$ and so there must be $b_1, \ldots, b_j \in |\mathscr{B}_l|$ such that if $\mu(\mathbf{x}_i) = b_i$ for $i = 1, \ldots, q$, then $\mathfrak{B} \models_l \Psi'_{h,k}[\mu]$. Then let $g(k)$ be the function with $\text{dom}(g(k)) = \text{rng}(h_1(k))$ and $g(k)(a_i) = b_i$ if $\bar{a}_i$ was replaced by $\mathbf{x}_i$ in constructing $\Psi'_{h,k}$. Now if $k \in \text{dom}(f) \cap \text{rng}(m)$, let $f(k) \in m^{-1}(k)$. Let $h_1(k) = \{a_0, \ldots, a_q\}$. Let $a_{i_1}, \ldots, a_{i_s}$ be the elements of $\{a_1, \ldots, a_q\} \cap \text{rng}(m)$ and let $a_{j_1}, \ldots, a_{j_t}$ be the elements of $\{a_1, \ldots, a_q\} - \text{rng}(m)$. Then

$$\mathfrak{A} \models_k (\exists \mathbf{x}_{j_1} \ldots \exists \mathbf{x}_{j_t} \Psi'_{h,k})_{\mathbf{x}_{i_1}, \ldots, \mathbf{x}_{i_s}} [a_{i_1}, \ldots, a_{i_s}]$$

and since m is elementary and $m(f(k)) = k$,

$$\mathfrak{B} \models_{f(k)} (\exists \mathbf{x}_{j_1} \ldots \exists \mathbf{x}_{j_t} \Psi'_{h,k})_{\mathbf{x}_{i_1}, \ldots, \mathbf{x}_{i_s}} [(m \upharpoonright \mathbf{B}_{f(k)})^{-1}(a_{i_1}), \ldots, (m \upharpoonright \mathbf{B}_{f(k)})^{-1}(a_{i_s})].$$

So let $g(k)(a_{i_\alpha}) = (m \upharpoonright |\mathscr{B}_{f(k)}|)^{-1}(a_{i_\alpha})$ for $\alpha = 1, \ldots, s$, and let $b_1, \ldots, b_s$ be such that

$$\mathfrak{B} \models_{f(k)} (\Psi'_{h,k})_{\mathbf{x}_{i_1}, \ldots, \mathbf{x}_{i_s} \mathbf{x}_{j_1}, \ldots, \mathbf{x}_{j_t}} [g(k)(a_{i_1}), \ldots, g(k)(a_{i_s}), b_1, \ldots, b_t].$$

Then let $g(k)(a_{j_\beta}) = b_\beta$ for $\beta = 1, \ldots, t$. It is now simple to compute that $\langle f, g \rangle \in J_h$.

Define J_k and $J_{k,\bar{a}}$ as in the proof of Theorem 6.1 and let

$$N = \{J_h : h \text{ is an } m\text{-formula bundle over } \mathfrak{A}\} \cup \{J_k : k \in \mathbf{K}\} \cup \{J_{k,\vec{a}} : k \in \mathbf{K} \,\&\, \vec{a} \in |\mathscr{A}_k|\}.$$

That N has the finite intersection property can be seen from the following example. Let h_1, h_2 be m-formula bundles over $\mathfrak{A}$, $k \in \mathbf{K}$ and $a_1, \ldots, a_s \in |\mathscr{A}_k|$. Define h as follows:

$$\text{dom}(h) = \text{dom}(h_1) \cup \text{dom}(h_2) \cup \{k\}.$$

For $k' \in \text{dom}(h)$, $k' \neq k$, let $h_1(k')$ first enumerate $\text{rng}(h_1^1(k'))$ and then $\text{rng}(h_1^2(k')) - \text{rng}(h_1^1(k'))$. Let $h_2(k') = h_2^1(k') \cup h_2^2(k')$ where if necessary, some free variables in underlying formulas in $h_2^2(k')$ are altered to give the right substitution. Let $h_1(k) = \{a_1, \ldots, a_s\}$ and $h_2(k) = \{\bar{a}_1 = \bar{a}_1, \ldots, \bar{a}_s = \bar{a}_s\}$. It is easy to see that $\emptyset \neq J_h \subseteq J_{h_1} \cap J_{h_2} \cap J_k \cap J_{k,\vec{a}}$.

So N can be extended to an ultrafilter F. Then $\bar{p}$ and p are defined just as $\bar{m}$ and m in the proof of Theorem 6.11, and the calculation that p is an elementary monomorphism is as given there. Let $g: \mathfrak{B} \to \mathfrak{B}^I_F$ be the canonical embedding.

We must show $\mathrm{p} \circ \mathrm{m} \approx g$. Let $l \in \mathbf{L}$. Then $g(l) = \langle l : i \in I \rangle / F$ and $(\mathrm{p} \circ \mathrm{m})(l) = \bar{p}(m(l))/F$. So $g(l) = (\mathrm{p} \circ \mathrm{m})(l)$ iff

$$C = \{\langle f, g \rangle \in I : \bar{p}(m(l))(\langle f, g \rangle) = l\} \in F.$$

If $l = \mathbf{P}$, $C = I \in F$. Suppose $l \neq \mathbf{P}$. Then

$$\begin{aligned} \bar{p}(m(l))(\langle f, g \rangle) = l \quad &\text{iff} \quad m(l) \in \mathrm{dom}(f) \,\&\, f(m(l)) = l \\ &\text{iff} \quad m(l) \in \mathrm{dom}(f). \end{aligned}$$

Thus $C = J_{m(l)} \in F$. Now let $b \in |\mathscr{B}_l|$, so that $g(b) = \langle b : i \in I \rangle / F$ and $(\mathrm{p} \circ \mathrm{m})(b) = \bar{p}(m(b))$. Again, $g(b) = (\mathrm{p} \circ \mathrm{m})(b)$ iff

$$D = \{\langle f, g \rangle \in I : \bar{p}(m(b))(\langle f, g \rangle) = b\} \in F.$$

If b is the fixed element used in the definition of $\bar{p}$, then $D = I \in F$. And if b is not this fixed element, then as above, $D = J_{m(l),\, m(b)} \in F$. Thus $g \approx p \circ m$. ■

We say that a class $K \subseteq \mathrm{St}(\mathrm{ML}, \Gamma)$ is an *elementary class* if there is a sentence **A** of ML such that K consists of all elements of $\mathrm{St}(\mathrm{ML}, \Gamma)$ in which **A** is valid; as in the classical case, we write $K \in \mathrm{EC}$ in this situation. We say that K is an *elementary class in the wider sense* ($K \in \mathrm{EC}_\Delta$) if there is a theory T formulated in ML such that K consists of all elements of $\mathrm{St}(\mathrm{ML}, \Gamma)$ which are models of T. The proofs of the following two theorems are virtually identical with those for the corresponding classical results as given in Bell and Slomsen (1969), pp. 151–152; we omit the details.

THEOREM 7.3. $K \in \mathrm{EC}_\Delta$ iff K is closed under ultraproducts and elementary equivalence.

THEOREM 7.4. $K \in \mathrm{EC}$ iff both K and $\mathrm{St}(\mathrm{ML}, \Gamma) - K$ are closed under ultraproducts and K is closed under elementary equivalence.

§8. DIRECT LIMITS

Except for §15, we will only have need to construct direct limits of systems indexed by the natural numbers. Moreover, due to our use of equivalence relations $\equiv_k$ to interpret equality in individual worlds, it is not clear how to extend our definition to more general direct systems. The proposed definition in Bowen (1975) fails to account for the presence of $\equiv_k$ in some applications below; however, whenever the direct system always satisfies $m_{i\zeta} = m_{ij} \circ m_{j\zeta}$, that definition works (and will be used in §15).

DEFINITION 8.1. A *direct system* $\{\mathfrak{A}_i, m_{ij}\}$ (*for* ML) consists of the following:

(a) for each i, a structure $\mathfrak{A}_i = \langle \mathscr{A}^i_k, \mathbf{K}^i, \mathbf{R}^i, \mathbf{O}^i, \mathbf{N}^i \rangle$ for ML;

(b) for all $i \leq j$, a monomorphism $m_{ij}: \mathfrak{A}_i \to \mathfrak{A}_j$ such that for all i, m_{ii} is the identity map and if $i \leq j \leq \zeta$, then we almost have $m_{i\zeta} \approx m_{ij} \circ m_{j\zeta}$ in the sense that if $k \in \mathbf{K}^i$, then $m_{i\zeta}(k) = (m_{ij} \circ m_{j\zeta})(k)$, and if $a \in |\mathscr{A}^i_k|$ and $k' = m_{i\zeta}(k)$, then

$$m_{i\zeta}(a) \equiv_{k'} (m_{ij} \circ m_{j\zeta})(a).$$

DEFINITION 8.2. Given a direct system $\{\mathfrak{A}_i, m_{ij}\}$ for ML, we define the *direct limit* $\varinjlim \mathfrak{A}_n$ or $\mathfrak{A}_\infty = \langle \mathscr{A}^\infty_k, \mathbf{K}^\infty, \mathbf{R}^\infty, \mathbf{O}^\infty, \mathbf{N}^\infty \rangle$ as follows. For each i, let $[i, \infty) = \{j : i \leq j\}$, and let $Z = \{[i, \infty) : i \geq 0\}$. Let

$$Z^* = \{f : \vee i \,\blacksquare\, \mathrm{dom}(f) = [i, \infty) \;\&\; \wedge j \geq i \,\blacksquare\, f(j) \in \mathbf{K}^j \;\& \\ f(j+1) = m_{j,\,j+1}(f(j))\}.$$

Given $f \in Z^*$, let δf be such that $\mathrm{dom}(f) = [\delta f, \infty)$. Given $f, g \in Z^*$, define $f \sim g$ iff for some $i \geq \max(\delta f, \delta g)$, $f(j) = g(j)$ for all $j \geq i$. It is easy to see that $f \sim g$ is an equivalence relation on Z^*. Let $[f]$ be the equivalence class of f under $\sim$ and let $\mathbf{K}^\infty = \{[f] : f \in Z^*\}$. Let $\mathbf{O}^\infty = [\lambda_i \,\blacksquare\, \mathbf{O}^i]$ and define $\mathbf{R}^\infty$ by

$$[f]\,\mathbf{R}^\infty [g] \quad \text{iff} \quad \vee i \wedge j \geq i [f(j)\, \mathbf{R}^j g(j)].$$

Clearly, $\mathbf{R}^\infty$ is defined independently of the choice of representatives. Similarly, let

$$\mathbf{N}^\infty([f]) \quad \text{iff} \quad \wedge i \geq \delta f\, \mathbf{N}^i(f(i)).$$

Next let $Z^{\#} = \{\langle f, g\rangle : f \in Z^* \,\&\, \mathrm{dom}(g) = \mathrm{dom}(f)$

$$\&\, \forall i \geq \delta f \,\blacksquare\, g(i) \in |\mathscr{A}^i_{f(i)}| \,\&\, g(i+1) = m_{i,\,i+1}(g(i)).$$

Given $\langle f, g\rangle, \langle f', g'\rangle \in Z^{\#}$, define

$$\langle f, g\rangle \sim \langle f', g'\rangle \quad \text{iff} \quad f \sim f' \,\&\, \vee i \wedge j \geq i[g(i) = g'(i)].$$

Again, this is an equivalence relation; let $[f, g]$ be the equivalence class of $\langle f, g\rangle \in Z^{\#}$. Given $\lfloor f \rfloor \in \mathbf{K}^{\infty}$, set

$$|\mathscr{A}^{\infty}_{\lfloor f \rfloor}| = \{[f', g] : f \sim f' \,\&\, \langle f', g\rangle \in Z^{\#}\}.$$

Given $[f_1', g_1], \ldots, [f_n', g_n] \in |\mathscr{A}^{\infty}_{\lfloor f \rfloor}|$ and $\mathbf{p}$, $\mathbf{f}$ in ML, define

$$\mathbf{p}^{\infty}_{\lfloor f \rfloor}([f_1', g_1], \ldots, [f_n', g_n]) \quad \text{iff} \quad \vee i \wedge j \geq i \mathbf{p}^i_{f(i)}(g_1(i), \ldots, g_n(i)),$$

and

$$\mathbf{f}^{\infty}_{\lfloor f \rfloor}([f'_1, g_1], \ldots, [f'_n, g_n]) = [h, g]$$

where δh is the least i such that for all $j \geq i$, $f(j) = f'_1(j) = \cdots = f'_n(j)$ and $h = f \upharpoonright [\delta h, \infty)$, and for $j \geq \delta h$, $g(j) = \mathbf{f}^j_{h(j)}(g_1(j), \ldots, g_n(j))$. Both of these definitions are independent of choice of representative. Similarly, if $[f, g], [f'', g'] \in |\mathscr{A}^{\infty}_{\lfloor f \rfloor}|$, define

$$[f', g] \equiv_{\lfloor f \rfloor} [f'', g'] \quad \text{iff} \quad \vee i \wedge j \geq i \,\blacksquare\, g(j) \equiv_{f(j)} g'(j).$$

This specifies $\mathfrak{A}_{\infty}$. Finally define $m_{i_{\infty}}$ from $\mathfrak{A}_i$ to $\mathfrak{A}_{\infty}$ by: for $z \in \mathrm{sk}(\mathfrak{A}_i)$, $m_{i_{\infty}}(z) = [f]$ where $\delta f = i$, $f(i) = z$, and for $j > i$, $f(j) = m_{j-1,\,j}(f(j-1))$. Then it is easy to verify that $m_{i_{\infty}}$ is a monomorphism, that for $i \leq j$, if $k \in \mathbf{K}^i$, then $m_{i_{\infty}}(k) = m_{j_{\infty}}(m_{ij}(k))$ and if $a \in |\mathscr{A}^i_k|$, then $m_{i_{\infty}}(a) \equiv_l m_{j_{\infty}}(m_{ij}(a))$, where $l = m_{i_{\infty}}(k)$. We will write $m_{i_{\infty}} \approx m_{j_{\infty}} \circ m_{ij}$ to indicate this state of affairs (cf. Bell and Slomsen (1971), pp. 165–167). Our direct limit also has the usual universal mapping property as follows: we omit the details of the proof (cf. Proposition 10.2 of Sacks (1972)).

THEOREM 8.3. Let $\{\mathfrak{A}_i, m_{ij}\}$ be a direct system of modal structures for ML and let $\mathfrak{B}$ be a modal structure for ML such that there are monomorphisms p_i from $\mathfrak{A}_i$ to $\mathfrak{B}$ which satisfy $p_j \circ m_{ij} \approx p_i$ when $i \leq j$. Then there exists a monomorphism q from $\mathfrak{A}_{\infty}$ to $\mathfrak{B}$ such that $q \circ m_{i_{\infty}} \approx p_i$ for each i.

LEMMA 8.4. Let S be one of the modal systems specified in §2 and let $\{\mathfrak{A}_i, m_{ij}\}$ be a direct system of S-structures. Then $\varinjlim \mathfrak{A}_i$ is again an S-structure.

Proof. It is easy to see that in the sense of classical first-order model theory, $b(\mathfrak{A}_\infty) \simeq \varinjlim(b(\mathfrak{A}_c))$. Now each of the sentences CD1–CD14 and Norm is equivalent to a $\forall\exists$-sentence of LB. But it is known (cf. Chang (1959) or Łos and Susko (1957)) that such sentences are preserved by direct limits. The result now follows. ■

THEOREM 8.5. Let $\{\mathfrak{A}_i, m_{ij}\}$ be a direct system of modal structures such that m_{ij} is an elementary embedding. Then each m_{i_∞} is an elementary embedding of $\mathfrak{A}_i$ in $\mathfrak{A}_\infty$.

Proof. We must show that

$$\mathfrak{A}_i \models_k \mathbf{A}[v] \quad \text{iff} \quad \mathfrak{A}_\infty \models_l \mathbf{A}[m_{i_\infty} \circ v],$$

where $l = m_{i_\infty}(k)$, $\mathbf{A}$ is any formula of ML, v is any assignment in $\mathfrak{A}_i$, and $k \in \mathbf{K}^i$. The proof proceeds by induction on the structure of $\mathbf{A}$. We will only consider the case in which $\mathbf{A}$ is $\Diamond\mathbf{B}$. Suppose that $\mathfrak{A}_i \models_k \Diamond\mathbf{B}[v]$. If $k \in \mathbf{K}^i - \mathbf{N}^i$, then by the definition of monomorphisms, $m_{ij}(k) \in \mathbf{K}^j - \mathbf{N}^j$ for all j with $i \leq j$, and so $m_{i_\infty}(k) \in \mathbf{K}^\infty - \mathbf{N}^\infty$. Thus we may assume that $k \in \mathbf{N}^i$. Then for some $k' \in \mathbf{K}^i$ with $k\mathbf{R}^i k'$, we have $\mathfrak{A}_i \models_{k'} \mathbf{B}[v]$. But then using induction, $m_{i_\infty}(k)\, \mathbf{R}^\infty m_{i_\infty}(k')$ and $\mathfrak{A}_\infty \models_{l'} \mathbf{B}[m_{i_\infty} \circ v]$, where $l' = m_{i_\infty}(k')$, and so $\mathfrak{A}_\infty \models_l \mathbf{A}[m_{i_\infty} \circ v]$. Now suppose that $\mathfrak{A}_\infty \models_l \Diamond\mathbf{B}[m_{i_\infty} \circ v]$. If $l \in \mathbf{K}^\infty - \mathbf{N}^\infty$, then for no j with $i \leq j$ do we have $m_{ij}(k) \in \mathbf{N}^j$; in particular then, $k = m_{ii}(k) \in \mathbf{K}^i - \mathbf{N}^i$. Thus we may assume that $l \in \mathbf{N}^\infty$ so that for some j with $i \leq j$, $m_{ij}(k) \in \mathbf{N}^j$. But since m_{ij} is a monomorphism, this implies that $k \in \mathbf{N}^c$. Now we must have $\mathfrak{A}_\infty \models_{l'} \mathbf{B}[m_{i_\infty} \circ v]$ for some l' with $l\mathbf{R}^\infty l'$. Then for some j with $i \leq j$ there is a $k' \in \mathbf{K}^d$ with $m_{j_\infty}(k') = l'$. Since $l\mathbf{R}^\infty l'$, there must be an e with $i \leq e$ and $j \leq e$ such that $m_{ie}(k)\mathbf{R}^e m_{je}(k')$. Moreover, since $m_{e_\infty}(m_{je}(k')) = l'$ and $m_{i_\infty} \circ v \approx (m_{e_r} \circ m_{ie}) \circ v$ it follows by induction that

$$\mathfrak{A}_e \models_{k''} \mathbf{B}[m_{ie} \circ v],$$

where $k'' = m_{je}(k')$, and so $\mathfrak{A}_e \models_{l''} \Diamond\mathbf{B}[m_{ie} \circ v]$, where $l'' = m_{ie}(k)$. But m_{ie} is an elementary embedding of $\mathfrak{A}_i$ in $\mathfrak{A}_e$, so it follows that $\mathfrak{A}_i \models_k \Diamond\mathbf{B}[v]$, as desired. ■

DEFINITION 8.6. Let $\{\mathfrak{A}_i, m_{ij}\}$ and $\{\mathfrak{B}_i, p_{ij}\}$ be direct systems of modal structures for ML, and for each i, let q_i be a monomorphism of $\mathfrak{A}_i$ into $\mathfrak{B}_i$ such that for $i \leq j$, $q_i \circ m_{ij} \approx p_{ij} \circ q_i$. We define the map $q_\infty = \varinjlim q_i$ from $\mathfrak{A}_\infty$ to $\mathfrak{B}_\infty$ as follows. If $[f] \in \mathbf{K}^\infty$, then $q_\infty([f]) = [\lambda i \centerdot q_i(f(i))]$, and if $[f', g] \in |\mathscr{A}^\infty_{[f]}|$, then $q_\infty([f', g]) = [\lambda i \centerdot (q_i(f'(i))), \lambda i \centerdot (q_i(g(i)))]$. That these specifications are independent of the choice of representatives is an easy computation. Moreover, $q_\infty \circ m_{i_\infty} \approx p_{i_\infty} \circ q_i$ for all i.

If for each n, $\langle I_n, F_n \rangle$ is an ultrafilter pair, then the sequence $\langle\langle I_n, F_n \rangle : n < \omega \rangle$ is an *ultrafilter sequence.*

DEFINITION 8.7. Let $U = \langle\langle I_n, F_n \rangle : n < \omega \rangle$ be a fixed ultrafilter sequence and let $\mathfrak{A}$ be a modal structure for ML. We define the *ultralimit of* $\mathfrak{A}$ *with respect to* U, $\mathfrak{A}^{\mathrm{U}}_{\infty}$, as follows. We first define a sequence $\langle \mathfrak{A}_n : n < \omega \rangle$ of modal structures for ML by induction: $\mathfrak{A}_0 = \mathfrak{A}$ and $\mathfrak{A}_{n+1} = (\mathfrak{A}_n)^{I_n}_{F_n}$. Then let d_{nn} be the identity map of $\mathfrak{A}_n$ and let $d_{n,\,n+1}$ be the canonical embedding of $\mathfrak{A}_n$ in $\mathfrak{A}_{n+1}$. For $n + 1 < m$, define d_{nm} by

$$d_{nm} = d_{m-1,\,m} \circ d_{m-2,\,m-1} \circ \cdots \circ d_{n,\,n+1}.$$

Then $\{\mathfrak{A}_n, d_{nm}\}$ is a direct system, and we set

$$\mathfrak{A}^{\mathrm{U}}_{\infty} = \varinjlim \mathfrak{A}_n.$$

By Theorems 6.5 and 8.5, $\mathfrak{A}$ is elementarily embeddable in $\mathfrak{A}^{\mathrm{U}}_{\infty}$ and so $\mathfrak{A} \equiv \mathfrak{A}^{\mathrm{U}}_{\infty}$. By Corollary 6.3 and Lemma 8.4, if S is one of the modal systems specified in §2 and $\mathfrak{A}$ is an S-structure, then $\mathfrak{A}^{\mathrm{U}}_{\infty}$ is also an S-structure. The next theorem extends a result of Kochen (1962).

THEOREM 8.8. Let $\mathfrak{A}$ and $\mathfrak{B}$ be modal structures for ML. Then $\mathfrak{A} \equiv \mathfrak{B}$ iff there exist ultralimits $\mathfrak{A}^{\mathrm{U}}_{\infty}$ and $\mathfrak{B}^{\mathrm{V}}_{\infty}$ together with an onto map $m : \mathfrak{A}^{\mathrm{U}}_{\infty} \xrightarrow{\equiv} \mathfrak{B}^{\mathrm{V}}_{\infty}$.

Proof. Sufficiency follows from the remarks above. For necessity, we adapt the proof of the classical version (cf. Bell and Slomsen (1971), p. 168). Let $\mathfrak{A}_0 = \mathfrak{A}$ and $\mathfrak{B}_0 = \mathfrak{B}$. By Theorem 7.1 there exist an ultrafilter pair $\langle I_0, F_0 \rangle$ and a map $h_0 : \mathfrak{B}_0 \xrightarrow{w\equiv} (\mathfrak{A}_0)^{I_0}_{F_0} = \mathfrak{A}_1$. Then by iterated

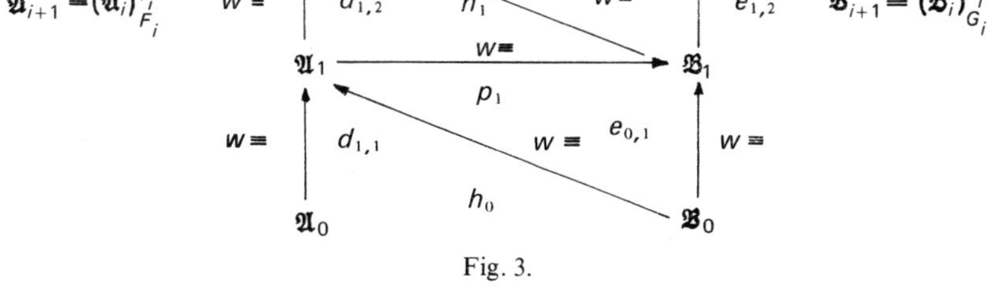

Fig. 3.

use of Theorem 7.2 we can erect an infinitely proceeding weakly commutative diagram (see Figure 3).

Then $m = \varinjlim p_i$ is the required map of $\mathfrak{A}^{\mathrm{U}}_{\infty}$ onto $\mathfrak{B}^{\mathrm{V}}_{\infty}$, where $\mathrm{U} = \langle\langle I_n, G_n\rangle : n < \omega\rangle$ and $\mathrm{V} = \langle\langle J_n, G_n\rangle : n < \omega\rangle$ (cf. Bowen (1975) for further details). ■

As in the classical case we immediately obtain the following corollary.

COROLLARY 8.9. Let S be one of the systems specified in §2 and let $K \subseteq \mathrm{St}(\mathrm{ML}, \Gamma_{\mathrm{S}})$. Then

(i) $K \in \mathrm{EC}_{\Delta}$ iff K is closed under isomorphism, ultraproducts and ultralimits and $\mathrm{St}(\mathrm{ML}, \Gamma_{\mathrm{S}}) - K$ is closed under ultralimits.

(ii) $K \in \mathrm{EC}$ iff both K and $\mathrm{St}(\mathrm{ML}, \Gamma_{\mathrm{S}}) - K$ are closed under isomorphism, ultraproducts and ultralimits.

§9. MODEL EXTENSIONS

Let $\mathfrak{A}$ and $\mathfrak{B}$ be modal structures for ML and let $m: \mathfrak{A} \to \mathfrak{B}$. If m is a Γ-morphism we say that $\mathfrak{B}$ is a *Γ-extension of* $\mathfrak{A}$ *via m*. A set Γ of formulas of ML is said to be *regular* if it includes all formulas of the forms $\mathbf{x} = \mathbf{y}$ and $\mathbf{x} \neq \mathbf{y}$ and also contains $\mathbf{A}_{\mathbf{x}_1 \ldots \mathbf{x}_n}[\mathbf{y}_1 \ldots \mathbf{y}_n]$ whenever it contains $\mathbf{A}$. A formula $\mathbf{A}$ of ML is said to be a *$**$-formula* if it is obtained from a formula of the form θ_h^* (cf. Definition 6.8), where h was a basic formula bundle over $\mathfrak{A}$, by replacing all names of elements of $|\mathscr{A}_\mathbf{O}|$ by free variables (so that $\wedge h_2(\mathbf{O})$ becomes a conjunction of basic formulas of ML).

MODEL EXTENSION THEOREM (9.1). Let S be a modal system which possesses a strongly characteristic set Γ_S. Let $\mathfrak{A}$ be an S-structure for ML which is a weak tree, let T be an S-theory, and let Γ be a regular set of formulas of ML which includes all $**$-formulas. Then $\mathfrak{A}$ possesses a Γ-extension $\mathfrak{B}$ via some (strong) faithful Γ-embedding m such that $\mathfrak{B}$ is model of T, if and only if, for all n and all $\mathbf{A}_1, \ldots, \mathbf{A}_n \in \Gamma$ if $\vdash_\mathrm{T}^\mathrm{S} \neg \mathbf{A}_1 \vee \ldots \vee \neg \mathbf{A}_n$, then $\neg \mathbf{A}_1 \vee \ldots \vee \neg \mathbf{A}_n$ is valid in $\mathfrak{A}$.

Proof. Suppose that $\mathfrak{B}$ is a Γ-extension of $\mathfrak{A}$ via m and that $\mathfrak{B}$ is a model of T. Let $\mathbf{A}_1, \ldots, \mathbf{A}_n \in \Gamma$, let v be an assignment in $\mathfrak{A}$, and suppose that $\vdash_\mathrm{T}^\mathrm{S} \neg \mathbf{A}_1 \vee \ldots \vee \neg \mathbf{A}_n$. Then by Definition 2.8, for some i, $1 \leq i \leq n$, $\mathfrak{B} \models_\mathbf{P} \neg \mathbf{A}[\mathrm{m} \circ v]$. But since $\mathbf{A}_i \in \Gamma$ and $\mathfrak{B}$ is a Γ-extension of $\mathfrak{A}$, we cannot have $\mathfrak{A} \models_\mathbf{O} \mathbf{A}_i[v]$. Consequently $\mathfrak{A} \models \neg \mathbf{A}_i[v]$ and it follows that $\neg \mathbf{A}_1 \vee \ldots \vee \neg \mathbf{A}_n$ is valid in $\mathfrak{A}$.

For the converse, assume the given condition on $\mathfrak{A}$, T, and Γ, let ML′ be the language obtained from ML by adding all the names of elements of $|\mathscr{A}_\mathbf{O}|$ as new individual constants, let U be the theory T as formulated in ML′, and let $\mathfrak{A}'$ be $(\mathfrak{A}, \underline{a})_{a \in |\mathscr{A}_\mathbf{O}|}$. Let V be the S-theory obtained from T by adding all the formulas of $\mathrm{D}\Gamma(\mathfrak{A}) \cup \mathrm{MD}^\Gamma(\mathfrak{A})$ as new nonlogical axioms, where $\mathrm{D}\Gamma(\mathfrak{A})$ consists of all sentences $\mathbf{A}$ of ML′ which are obtained from a formula of Γ by replacing all free variables by names of elements of $|\mathscr{A}_\mathbf{O}|$ and such that $\mathfrak{A}' \models_\mathbf{O} \mathbf{A}$. Suppose V were inconsistent. Then there would be $\mathbf{A}_1', \ldots, \mathbf{A}_n' \in \mathrm{D}\Gamma(\mathfrak{A})$ and $\mathbf{B}_1', \ldots, \mathbf{B}_p' \in \mathrm{MD}^\Gamma(\mathfrak{A})$ such that

$$(9.2) \qquad \vdash_\mathrm{U}^\mathrm{S} \neg \mathbf{A}_1' \vee \ldots \vee \neg \mathbf{A}_n' \vee \neg \mathbf{B}_1' \vee \ldots \vee \neg \mathbf{B}_p'.$$

Let $\mathbf{e}_1, \ldots, \mathbf{e}_t$ be all the new individual constants of ML′ which occur in the formula of (9.2) and let $\mathbf{x}_1, \ldots, \mathbf{x}_t$ be distinct new variables. Then by the Theorem on Constants (1.6),

$$\vdash^{\mathrm{S}}_{\mathrm{T}} \neg \mathbf{A}_1 \vee \ldots \vee \neg \mathbf{A}_n \vee \neg \mathbf{B}_1 \vee \ldots \vee \neg \mathbf{B}_n,$$

where $\mathbf{A}_i$ (resp. $\mathbf{B}_i$) results from $\mathbf{A}'_i$ (resp. $\mathbf{B}'_i$) by replacing $\mathbf{e}_j$ everywhere by $\mathbf{x}_j$ for $j = 1, \ldots, t$. Since $\mathbf{B}_i$ is a $**$-formula, $\mathbf{B}_i \in \Gamma$, and then by the regularity of Γ, $\mathbf{A}_1, \ldots, \mathbf{A}_n, \mathbf{B}_1, \ldots, \mathbf{B}_p$ all belong to Γ. Hence by hypothesis, the formula

$$(9.3) \qquad \neg \mathbf{A}_1 \vee \ldots \vee \neg \mathbf{A}_n \vee \neg \mathbf{B}_1 \vee \ldots \vee \neg \mathbf{B}_p$$

is valid in $\mathfrak{A}$. But let v be an assignment in $\mathfrak{A}$ such that for $\alpha = 1, \ldots, t$, $v(\mathbf{x}_\alpha)$ is named by $\mathbf{e}_\alpha$. Then by definition of $\mathrm{D}\Gamma(\mathfrak{A})$ and $\mathrm{MD}(\mathfrak{A})$,

$$\mathfrak{A} \models \mathbf{A}_i[v] \quad \text{and} \quad \mathfrak{A} \models \mathbf{B}_j[v]$$

for $i = 1, \ldots, n$ and $j = 1, \ldots, p$, a contradiction. Thus V must be S-consistent. Since Γ_{S} is strongly characteristic for S, there exists an S-structure $\mathfrak{C}$ for ML′ which is a model of V. Since $\mathfrak{C}$ is a model of $\mathrm{MD}^{\Gamma}(\mathfrak{A})$, by Theorem 6.11 there exists an ultrapower $\mathfrak{B}$ of $\mathfrak{C}$ together with a Γ-embedding m of $\mathfrak{A}$ in $\mathfrak{B}$. Since $\mathfrak{C}$ is a model of T and $\mathfrak{C} \prec \mathfrak{B}$, then $\mathfrak{B}$ is a model of T as desired. Note that we can get m to be strong or faithful by using Theorem 6.13 instead of Theorem 6.11 above. ■

THEOREM 9.4. Let S possess a strongly characteristic set Γ_{S} and let $\mathfrak{A}$ and $\mathfrak{B}$ be S-structures for ML which are weak trees. Let Γ be a regular set of formulas of ML which includes all $**$-formulas, let Δ include all sentences of the form

$$\Box^{\alpha} \forall \mathbf{x}_1 \ldots \forall \mathbf{x}_m [\neg \mathbf{A}_1 \vee \ldots \vee \neg \mathbf{A}_n]$$

where $\mathbf{A}_1, \ldots, \mathbf{A}_n \in \Gamma$ and $\alpha \geq 0$, and let $m: \mathfrak{A} \xrightarrow{\mathrm{S},\Delta} \mathfrak{B}$ be a faithful strong Δ-embedding. Then there exists an S-modal structure $\mathfrak{C}$ and faithful maps $m': \mathfrak{B} \xrightarrow{\mathrm{S},\Gamma} \mathfrak{C}$ and $m'': \mathfrak{A} \xrightarrow{\equiv} \mathfrak{C}$ such that $m'' \approx m' \circ m$, so that Figure 4 weakly commutes.

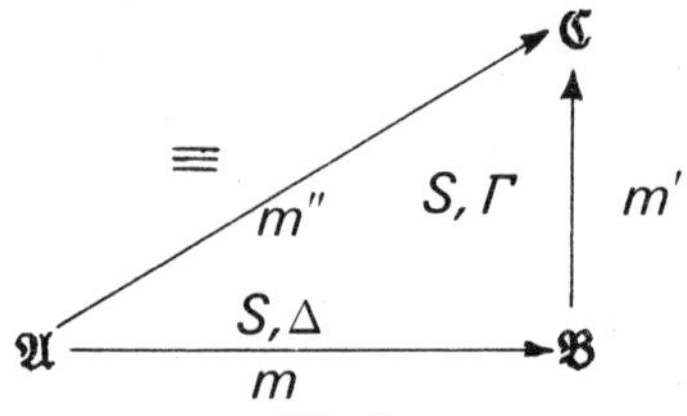

Fig. 4.

Proof. Let h be any Γ-formula bundle over $\mathfrak{B}$. Let $l \in \mathrm{dom}(h)$ and let $\mathbf{A}_{\mathbf{x}_1 \cdots \mathbf{x}_n}[\underline{b}_1, \ldots, \underline{b}_n] \in h_2(l)$. First suppose that $k \in \mathbf{K}$, that $m(k) = l$, and that

(9.5) $\quad \mathfrak{A} \models_k \neg \exists \mathbf{x}_1 \ldots \mathbf{x}_n \mathbf{A}.$

Then since m is a Δ-morphism, it follows that $\mathfrak{B} \models_l \forall \mathbf{x}_1 \ldots \mathbf{x}_n \neg \mathbf{A}$, a contradiction. Thus we must have that $\mathfrak{A} \models_k \exists \mathbf{x}_1 \ldots \mathbf{x}_n \mathbf{A}$. Now suppose that $l \notin \mathrm{rng}(m)$ and let t be such that $\mathbf{PS}^t l$. Suppose that

$$\mathfrak{A} \models_{\mathbf{O}} \neg \Diamond^t \exists \mathbf{x}_1 \ldots \mathbf{x}_n \mathbf{A}.$$

Again since m is a Δ-morphism, $\mathfrak{B} \models_{\mathbf{P}} \Box^t \forall \mathbf{x}_1 \ldots \mathbf{x}_n \neg \mathbf{A}$, and so $\mathfrak{B} \models_l \forall \mathbf{x}_1 \ldots \mathbf{x}_n \neg \mathbf{A}$, once again a contradiction. Thus there must be a $k \in \mathbf{K}$ with $\mathbf{OR}^t k$ and $\mathfrak{A} \models_k \exists \mathbf{x}_1 \ldots \mathbf{x}_n \mathbf{A}$.

Now let $\mathfrak{A}'$ be an $\aleph_0$-trivial expansion of $\mathfrak{A}$, and let I be the collection of m-bundles $\langle f, g \rangle$ from $\mathfrak{B}$ to $\mathfrak{A}'$ where f is $1-1$. It follows from the observations above that if we define J_h as in the proof of Theorem 7.2, then $J_h \neq \emptyset$. Now obtain F as in the proof of Theorem 7.2, and define m' as p is defined there. Let m^* be the identity embedding of $\mathfrak{A}$ in $\mathfrak{A}'$, and let $m^{\#}$ be the canonical embedding of $\mathfrak{A}'$ in $(\mathfrak{A}')^I_F$. Let $k \in \mathbf{K}$. We claim that $(m' \circ m)(k) = (m^{\#} \circ m^*)(k)$. Set $Z = \{\langle f, g \rangle \in I : m(k) \in \mathrm{dom}(f)\}$. As in earlier proofs, $Z \in F$. Let $\langle f, g \rangle \in I$, and note that $m^*(k) = k$ and $\overline{m}^{\#}(m^*(k))(\langle f, g \rangle) = k$. But since m is strong and $\langle f, g \rangle \in Z$, then

$$\overline{m}'(m(k))(\langle f, g \rangle) = f(m(k)) = k,$$

and the claim follows. Arguing similarly for individuals, it follows that $m' \circ m \approx m^{\#} \circ m^*$. Thus setting $\mathfrak{C} = (\mathfrak{A}')^I_F$ and $m'' = m^{\#} \circ m^*$, the theorem follows. ∎

DEFINITION 9.6. A formula $\mathbf{A}$ of ML is a $\Diamond\exists$-*formula* if it is built up from basic formulas of ML by means of $\wedge$, $\vee$, $\exists$, and $\Diamond$. A formula $\mathbf{A}$ of ML is a $\Box\forall$-formula if it is built up from basic formulas of ML by means of $\wedge$, $\vee$, $\forall$, and $\Box$.

We say that two S-theories T and T′ are *equivalent* if for every formula $\mathbf{A}$ of ML, $\vdash^{\mathrm{S}}_{\mathrm{T}} \mathbf{A}$ iff $\vdash^{\mathrm{S}}_{\mathrm{T}'} \mathbf{A}$.

THEOREM 9.7. Let S be a modal system having the weak tree property and possessing a strongly characteristic set Γ_{S} for which it is provable that for every $\Diamond\exists$-formula $\mathbf{A}$, there is a $\Box\forall$-formula $\mathbf{B}$ st $\vdash^{\mathrm{S}} \neg \mathbf{A} \leftrightarrow \mathbf{B}$ or $\vdash^{\mathrm{S}} \neg \mathbf{A} \leftrightarrow (\Box \forall x [x = x] \rightarrow \mathbf{B})$. An S-theory T is equivalent to some S-theory T′ all of whose nonlogical axioms are of the form $\mathbf{A}$ or $\forall x [x = x] \rightarrow \mathbf{A}$ where $\mathbf{A}$ is a $\Box\forall$-formula iff for every pair of S-structures

$\mathfrak{A}$ and $\mathfrak{B}$, if $m: \mathfrak{A} \to \mathfrak{B}$ and $\mathfrak{B}$ is a model of T, then $\mathfrak{A}$ is a model of T.

Proof. It is a routine computation to show that if $\mathbf{A}$ is a $\Box\forall$-formula and $m: \mathfrak{A} \to \mathfrak{B}$, then $\mathfrak{B} \models_{m(k)} \mathbf{A}[m \circ v]$ implies that $\mathfrak{A} \models_k \mathbf{A}[v]$ and $\mathfrak{B} \models_{m(k)} \forall x[x = x] \to \mathbf{A}[m \circ v]$ implies that $\mathfrak{A} \models_k \Box\forall x[x = x] \to \mathbf{A}[v]$. From this it is easy to prove the necessity of the condition.

Conversely, suppose the given condition holds and let W be the theory in ML whose nonlogical axioms consist of the $\Box\forall$-formulas $\mathbf{A}$ such that $\vdash^{\mathrm{S}}_{\mathrm{T}} \mathbf{A}$ together with the formulas $\Box\forall x[x = x] \to \mathbf{A}$ such that $\mathbf{A}$ is a $\Box\forall$-formula and $\vdash^{\mathrm{S}}_{\mathrm{T}} \Box\forall x[x = x] \to \mathbf{A}$. Let $\mathfrak{A}$ be an S-structure which is a weak tree and which is a model of W, and let Γ consist of all $\Diamond$E-formulas together with the formula $\Box\forall x[x = x]$. Then Γ is regular and includes all $**$-formulas. Let $\mathbf{A}_1, \ldots, \mathbf{A}_n \in \Gamma$ and suppose that $\vdash^{\mathrm{S}}_{\mathrm{T}} \neg\mathbf{A}_1 \vee \ldots \vee \neg\mathbf{A}_n$. Now $\neg\mathbf{A}_1 \vee \ldots \vee \neg\mathbf{A}_n$ is equivalent in S to a formula $\mathbf{B}$ which is either a $\Box\forall$-formula or of the form $\Box\forall x[x = x] \to \mathbf{A}$, where $\mathbf{A}$ is a $\Box\forall$-formula. Since $\vdash^{\mathrm{S}}_{\mathrm{T}} \mathbf{B}$, then $\mathbf{B}$ is a nonlogical axiom of W and hence is valid in $\mathfrak{A}$. Hence $\neg\mathbf{A}_1 \vee \ldots \vee \neg\mathbf{A}_n$ is valid in $\mathfrak{A}$. Then by the Model Extension Theorem there is a Γ-extension $\mathfrak{B}$ of $\mathfrak{A}$ which is a model of T. Then by the given condition, $\mathfrak{A}$ is a model of T. Hence every model of W is a model of T and so by the assumption that S is strongly complete with respect to Γ_{S}, it follows that T and W are equivalent in S. ■

A question which naturally arises is whether or not the Γ-extension $\mathfrak{B}$ of $\mathfrak{A}$ which the Model Extension Theorem provides can be assumed to be such that $b(\mathfrak{B}) \simeq b(\mathfrak{A})$. The following simple example shows that this is not possible. Let $\mathfrak{A} = \langle \mathscr{A}_k, \mathbf{K}, \mathbf{R}, \mathbf{O} \rangle$ be such that $\mathbf{K} = \{\mathbf{O}\}$, $\mathbf{R} = \{\langle \mathbf{O}, \mathbf{O} \rangle\}$, and $\mathscr{A}_{\mathbf{O}} = \langle \{a\}, p_{\mathbf{O}} \rangle$, where p is a unary predicate symbol and $p_{\mathbf{O}} = \{a\}$. Then $\mathscr{A}_{\mathbf{O}} \models \forall x p(x)$. Let T be the theory whose only axiom is $\Diamond\forall x \neg p(x)$, and let Γ be the set of all atomic formulas; i.e., Γ consists of all the formulas $p(\mathbf{y})$ as $\mathbf{y}$ ranges over all variables. Let $\mathfrak{B} = \langle \mathscr{B}_l, \mathbf{L}, \mathbf{S}, \mathbf{P} \rangle$ be the structure defined as follows (take $\mathbf{N} = \emptyset$): $\mathbf{L} = \{\mathbf{O}, \zeta\}$, $\mathbf{S} = \{\langle \mathbf{O}, \zeta \rangle, \langle \mathbf{O}, \mathbf{O} \rangle, \langle \zeta, \zeta \rangle\}$, $\mathbf{P} = \mathbf{O}$, $\mathscr{B}_{\mathbf{O}} = \mathscr{A}_{\mathbf{O}}$, and $\mathscr{B}_\zeta = \langle \{a\}, p_\zeta \rangle$, $p_\zeta = \emptyset$. Then $\mathscr{B}_\zeta \models \forall x \neg p(x)$, and so $\mathfrak{B} \models_{\mathbf{P}} \Diamond\forall x \neg p(x)$. Thus $\mathfrak{B}$ is a model of T. But since $\mathscr{B}_{\mathbf{P}} = \mathscr{A}_{\mathbf{O}}$ and Γ consists only of atomic formulas, $\mathfrak{B}$ is a Γ-extension of $\mathfrak{A}$. Hence $\mathfrak{A}$ possesses a Γ-extension which is a model of T. On the other hand, suppose that $\mathfrak{C}$ is a Γ-extension which is a model of T and which satisfies $b(\mathfrak{C}) = b(\mathfrak{A})$, say $\mathfrak{C} = \langle \mathscr{C}_m, \mathbf{K}, \mathbf{R}, \mathbf{O} \rangle$. Then $\mathfrak{A} \models_{\mathbf{O}} p(x)[a]$ implies $\mathfrak{C} \models_{\mathbf{O}} p(x)[a]$, so $\mathfrak{C} \models_{\mathbf{O}} \exists x p(x)$, and since the only element of $\mathbf{K}$ is $\mathbf{O}$ and $\mathbf{ORO}$, then $\mathfrak{C} \models_{\mathbf{O}} \Box\exists x p(x)$. But since $\mathfrak{C}$ is a model of T, we must have $\mathfrak{C} \models_{\mathbf{O}} \Diamond\forall x \neg p(x)$, a contradiction. Thus $\mathfrak{A}$ can possess no Γ-extension with the same base which is also a model of T.

COROLLARY 9.8. Let S be as in Theorem 9.7. A formula **A** is equivalent to a □∀-formula or to a formula of the form $\Box\forall\mathbf{x}[\mathbf{x}=\mathbf{x}]\rightarrow\mathbf{B}$ where **B** is a □∀-formula if and only if for all structures $\mathfrak{A}$ and $\mathfrak{B}$, all $m: \mathfrak{A}\rightarrow\mathfrak{B}$, and all $k\in\mathbf{K}$, if $\mathfrak{B}\models_{m(k)}\mathbf{A}[m\circ v]$, then $\mathfrak{A}\models_k\mathbf{A}[v]$, where v is any assignment in $\mathfrak{A}$.

Proof. One direction is a simple computation. For the other direction assume the condition and let T be the theory, whose only axiom is **A**. Then by Theorem 9.7, there is a theory T′ equivalent to T all of whose axioms are either □∀-formulas or of the form $\Box\forall\mathbf{x}[\mathbf{x}=\mathbf{x}]\rightarrow\mathbf{B}$ where **B** is a □∀-formula. Then $\vdash^{\mathrm{S}}_{\mathrm{T'}}\mathbf{A}$, and so there are axioms $\mathbf{C}_1,\ldots,\mathbf{C}_n$ of T′ such that

$$\vdash^{\mathrm{S}}\mathbf{C}_1\wedge\ldots\wedge\mathbf{C}_n\rightarrow\mathbf{A}.$$

But $\vdash^{\mathrm{S}}_{\mathrm{T}}\mathbf{C}_i$ for $i=1,\ldots,n$, so $\vdash^{\mathrm{S}}\mathbf{A}\rightarrow\mathbf{C}_i$, and hence

$$\vdash^{\mathrm{S}}\mathbf{A}\leftrightarrow(\mathbf{C}_1\wedge\ldots\wedge\mathbf{C}_n).$$

But $\mathbf{C}_1\wedge\ldots\wedge\mathbf{C}_n$ is equivalent to a formula of the required form, and so the result follows. ■

§10. INDUCTIVE THEORIES

We say that an S-modal theory T is *inductive* if for each direct system $\{\mathfrak{A}_n, f_n^m\}$ where $f_n^m: \mathfrak{A}_n \to \mathfrak{A}_m$ for $n \le m$, if each $\mathfrak{A}_n$ is a model of T, then $\varinjlim \mathfrak{A}_n$ is again a model of T. Recall that a formula is basic if it is either atomic or the negation of an atomic formula.

LEMMA 10.1. Let $\mathbf{A}$ be a $\Diamond$E-formula, let $\mathfrak{A}$ and $\mathfrak{B}$ be S-modal structures, let $m: \mathfrak{A} \to \mathfrak{B}$, and let $k \in \mathbf{K}$. Then for any assignment v in $\mathfrak{A}$, $\mathfrak{A} \models_k \mathbf{A}[v]$ implies that $\mathfrak{B} \models_{m(k)} \mathbf{A}[m \circ v]$.

Proof. By induction on the number of logical symbols in $\mathbf{A}$. For atomic $\mathbf{A}$ this follows from the definition of monomorphism, while for $\mathbf{A}$ of the forms $\mathbf{B} \vee \mathbf{C}$ and $\mathbf{B} \wedge \mathbf{C}$, the induction is easy. Suppose $\mathbf{A}$ is $\exists \mathbf{x} \mathbf{B}$ and $\mathfrak{A} \models_k \exists \mathbf{x} \mathbf{B}[v]$, so that for some $a \in |\mathscr{A}_k|$, $\mathfrak{A} \models_k \mathbf{B}[v(\substack{\mathbf{x}\\a})]$. Then by induction, $\mathfrak{B} \models_{m(k)} \mathbf{B}[m \circ (v(\substack{\mathbf{x}\\a}))]$. Now $m \circ (v(\substack{\mathbf{x}\\a})) = (m \circ v)(\substack{\mathbf{x}\\m(a)})$ and $m(a) \in |\mathscr{B}_{m(k)}|$, and it follows that $\mathfrak{B} \models_{m(k)} \exists \mathbf{x} \mathbf{B}[m \circ v]$. Finally, suppose that $\mathbf{A}$ is $\Diamond \mathbf{B}$ and $\mathfrak{A} \models_k \Diamond \mathbf{B}[v]$. If $k \in \mathbf{K} - \mathbf{N}$ then $m(k) \in \mathbf{L} - \mathbf{M}$, and so $\mathfrak{B} \models_{m(k)} \Diamond \mathbf{B}[m \circ v]$. If $k \in \mathbf{N}$, then $m(k) \in \mathbf{L}$, and for some $k' \in \mathbf{K}$ with $k \mathbf{R} k'$ we have $\mathfrak{A} \models_{k'} \mathbf{B}[v]$. Then using the induction hypothesis, $\mathfrak{B} \models_{m(k')} \mathbf{B}[m \circ v]$. Moreover, $m(k') \in \mathbf{L}$ and $m(k) \mathbf{S} m(k')$. It follows that $\mathfrak{B} \models_{m(k)} \Diamond \mathbf{B}[m \circ v]$, as desired. ∎

LEMMA 10.2. Let S be a modal system. Let T be an S-theory all of whose nonlogical axioms are $\Box\forall$-$\Diamond\exists$-formulas and let $\{\mathfrak{A}_n, f_n^m\}$ be a direct system of modal structures. Let $\mathbf{A}$ be a nonlogical axiom of T, let $n < \omega$, let v be an assignment in $\mathfrak{A}_n = \langle \mathscr{A}_k^n, \mathbf{K}^n, \mathbf{R}^n, \mathbf{O}^n, \mathbf{N}^n \rangle$, let $k \in \mathbf{K}^n$, and suppose that for all $m \ge n$ we have

$$\mathfrak{A}_m \models_l \mathbf{A}[f_n^m \circ v];$$

where $l = f_n^m(k)$. Then

$$\mathfrak{A}_\infty \models_l \mathbf{A}[f_n^\infty \circ v],$$

where $l = f_n^\infty(k)$.

Proof. We proceed by induction on the structure of $\mathbf{A}$. If $\mathbf{A}$ is a $\Diamond\exists$-formula, the result is immediate by Lemma 10.3, while if $\mathbf{A}$ is $\mathbf{B} \wedge \mathbf{C}$, the inductive procedure is obvious. Let $\mathbf{A}$ be $\mathbf{B} \vee \mathbf{C}$. Then we can construct a monotonically increasing, infinite sequence $\langle n_i : i < \omega \rangle$ such that $n \leq n_0$ and for all $i < \omega$, if $l = f_n^{n_i}(k)$, then

$$\mathfrak{A}_{n_i} \models_l \mathbf{B}[f_n^{n_i} \circ v].$$

(For if not, we deal with the formula $\mathbf{C} \vee \mathbf{B}$.) But $\mathfrak{A}_\infty = \varinjlim \mathfrak{A}_{n_i}$, and so by induction, $\mathfrak{A}_\infty \models_l \mathbf{B}[f_n^\infty \circ v]$ where $l = f_n^\infty(k)$, and so $\mathfrak{A}_\infty \models_l \mathbf{A}[f_n^\infty \circ v]$. If $\mathbf{A}$ is $\forall \mathbf{x}_1 \ldots \forall \mathbf{x}_m \mathbf{B}$, let $\bar{a}_1, \ldots, \bar{a}_m \in |A_l^\infty|$ where $l = f_n^\infty(k)$. Choose $\bar{n} \geq n$ so that there are $a_1, \ldots, a_m \in |A_l^{\bar{n}}|$ where $l = f_n^{\bar{n}}(k)$ such that $f_{\bar{n}}^\infty(a_i) = \bar{a}_i$ for $i = 1, \ldots, m$. By hypothesis, if $p \geq \bar{n}$, $\mathfrak{A}_p \models_l \mathbf{A}[f_n^p \circ v]$, where $l = f_n^p(k)$, so if

$$\mu_p = (f_n^p \circ v)\begin{pmatrix} a_1, \ldots, a_m \\ \mathbf{x}_1, \ldots, \mathbf{x}_m \end{pmatrix},$$

then $\mathfrak{A}_p \models_l \mathbf{B}[\mu_p]$. By induction, $\mathfrak{A}_\infty \models_l \mathbf{B}[f_n^\infty \circ \mu_{\bar{n}}]$, and so, since $f_{\bar{n}}^\infty \circ \mu_{\bar{n}}$ agrees with $f_n^\infty \circ v$ on the free variables of $\mathbf{A}$ (which were assumed distinct from $\mathbf{x}_1, \ldots, \mathbf{x}_m$), we have $\mathfrak{A}_\infty \models_l \mathbf{A}[f_n^\infty \circ v]$. Finally let $\mathbf{A}$ be $\Box\mathbf{B}$. It is easy to see from Definitions 8.2 and 2.11 that if $l \in \mathbf{K}^\infty - \mathbf{N}^\infty$, then for all n and k, if $f_n^\infty(k) = l$, then $k \in \mathbf{K}^n - \mathbf{N}^n$. Hence if $k \in \mathbf{N}^n$, then $l \in \mathbf{N}^\infty$. Assume $k \in \mathbf{N}^n$. Let $\bar{l} \in \mathbf{K}^\infty$ with $l\mathbf{R}^\infty \bar{l}$, where $l = f_n^\infty(k)$. Choose $\bar{n} \geq n$ so that for some $k' \in \mathbf{K}^{\bar{n}}$, $f_{\bar{n}}^\infty(k') = \bar{l}$. By hypothesis we have $\mathfrak{A}_p \models_{l'} \Box\mathbf{B}[f_n^p \circ v]$ whenever $p \geq \bar{n}$ and $l' = f_n^p(k)$, and so $\mathfrak{A}_p \models_{l''} \mathbf{B}[f_n^p \circ v]$ where $l'' = f_n^p(k')$. It follows by induction that $\mathfrak{A}_\infty \models_{\bar{l}} \mathbf{B}[f_n^\infty \circ v]$ and so $\mathfrak{A}_\infty \models_l \mathbf{A}[f_n^\infty \circ v]$. ■

COROLLARY 10.3. Let T be an S-theory all of whose nonlogical axioms are $\Box\forall$-$\Diamond\exists$-formulas. Then T is inductive.

Proof. Let $\{\mathfrak{A}_n, f_n^m\}$ be a direct system of S-modal models of T and let $\mathbf{A}$ be a nonlogical axiom of T with free variables $\mathbf{x}_1, \ldots, \mathbf{x}_n$. Let v be an assignment in $\mathfrak{A}_\infty$ and choose n sufficiently large so that there are $a_1, \ldots, a_n \in \mathrm{U}(\mathfrak{A}_n)$ so that $f_n^\infty(a_i) = v(\mathbf{x}_i)$, $i = 1, \ldots, n$. Let μ be an assignment in $\mathfrak{A}_n$ so that $\mu(\mathbf{x}_i) = a_i$ for $i = 1, \ldots, n$. Then by hypothesis, for $m \geq n$ we have $\mathfrak{A}_m \models \mathbf{A}[f_n^m \circ \mu]$, and so by Lemma 10.4, $\mathfrak{A}_\infty \models \mathbf{A}[f_n^\infty \circ \mu]$. Since $(f_n^\infty \circ \mu)(\mathbf{x}_i) \equiv_{\mathbf{0}^\infty} v(x_i)$ for $i = 1, \ldots, n$, it follows that $\mathfrak{A}_\infty \models \mathbf{A}[v]$ and so $\mathfrak{A}_\infty$ is a model of T. ■

THEOREM 10.6. Let S be a modal system having the weak tree property which possesses a strongly characteristic set Γ_S and is such that in S, every $**$-formula is equivalent to the negation of some $\Box\forall$-$\Diamond\exists$-formula. An S-theory T is inductive iff T is equivalent in S to an S-theory W all of whose nonlogical axioms are $\Box\forall$-$\Diamond\exists$-formulas.

Proof. One direction is immediate from Corollary 10.5. For the other direction let W be the S-theory whose nonlogical axioms consist of all $\Box\forall$-$\Diamond\exists$-formulas $\mathbf{A}$ such that $\vdash^{\mathrm{S}}_{\mathrm{T}}\mathbf{A}$, and let Γ be the set of formulas which are equivalent in S to the negation of a $\Box\forall$-$\Diamond\exists$-formula. Clearly Γ is regular. Since both $\Box\forall x[x = x]$ and $\Diamond\exists x[x \neq x]$ are $\Box\forall$-$\Diamond\exists$-formulas, it follows from remarks above that every $**$-formula belongs to Γ. Let $\mathfrak{A}$ be an S-structure which is a model of W. We will define a direct system $\{\mathfrak{A}_n, f_n^m\}$ such that $\mathfrak{A}_1 = \mathfrak{A}$, $f_n^m : \mathfrak{A}_n \xrightarrow{s} \mathfrak{A}_m$ for $n \leq m$, and for each n, $\mathfrak{A}_{2n}$ is a model of T, $f_1^{2n+1} : \mathfrak{A}_1^s \xrightarrow{\equiv} \mathfrak{A}_{2n+1}$, and $f_{2n+1}^{2n+2} : \mathfrak{A}_{2n+1} \xrightarrow{\Gamma} \mathfrak{A}_{2n+2}$. For $\mathfrak{A}_2$, assume that $\mathbf{B}_1, \ldots, \mathbf{B}_p \in \Gamma$ and $\vdash^{\mathrm{S}}_{\mathrm{T}} \neg \mathbf{B}_1 \vee \ldots \vee \neg \mathbf{B}_p$. Let $\mathbf{A}_1, \ldots, \mathbf{A}_p$ be $\Box\forall$-$\Diamond\exists$-formulas such that $\vdash^{\mathrm{S}}_{\mathrm{T}}\mathbf{B}_i \leftrightarrow \neg \mathbf{A}_i$ for $i = 1, \ldots, p$. Let $\mathbf{D}$ be $\mathbf{A}_1 \vee \ldots \vee \mathbf{A}_p$. Then $\vdash^{\mathrm{S}}_{\mathrm{T}}\mathbf{D}$. Since $\mathbf{D}$ is a $\Box\forall$-$\Diamond\exists$-formula, then $\mathbf{D}$ is an axiom of W and hence valid in $\mathfrak{A}$. If $\mathbf{D}'$ is the universal closure of D, then $\mathfrak{A} \models \mathbf{D}'$. Hence $\neg \mathbf{B}_1 \vee \ldots \vee \neg \mathbf{B}_p$ is valid in $\mathfrak{A}_1$. Consequently, by the Model Extension Theorem there is a Γ-extension $\mathfrak{A}_2$ of $\mathfrak{A}_1$ via a strong Γ-morphism f_1^2 which is a S-model of T.

Now suppose that $\mathfrak{A}_k, f_k^{k'}$ have been defined as desired for all $k, k' \leq 2n$, $n \geq 1$. To define $\mathfrak{A}_{2n+1}$, let Π consist of all $\Diamond\exists$-formulas together with $\Box\forall\mathbf{x}[\mathbf{x} = \mathbf{x}]$ and all $**$-formulas so that Π is regular and includes all $**$-formulas. Let Δ consist of all formulas of the form $\Box^{\mathrm{S}}\forall \mathbf{x}_1 \ldots \mathbf{x}_n \mathbf{A}$, where $\mathbf{A}$ is a disjunction of negations of formulas in Π. Then it is easy to see that any formula in Δ is equivalent to one of the two forms $\neg\mathbf{B}$ or $\neg\mathbf{B} \vee \neg\Box\forall\mathbf{x}[\mathbf{x} = \mathbf{x}]$, where $\mathbf{B}$ is a $\Diamond\exists$-formula, and consequently, $\Delta \subseteq \Gamma$. By induction, f_{2n-1}^{2n} is a strong Γ-embedding of $\mathfrak{A}_{2n-1}$ in $\mathfrak{A}_{2n}$, and hence f_{2n-1}^{2n} is a strong Δ-embedding of $\mathfrak{A}_{2n-1}$ in $\mathfrak{A}_{2n}$. Hence by Theorem 9.4, there exists an S-structure $\mathfrak{A}_{2n+1}$ together with maps $f_{2n}^{2+1} : \mathfrak{A}_{2n} \xrightarrow{s, \Pi} \mathfrak{A}_{2n+1}$ and $f_{2n-1}^{2n} : \mathfrak{A}_{2n-1} \xrightarrow{\Pi} \mathfrak{A}_{2n+1}$ such that $f_{2n-1}^{2n+1} = f_{2n}^{2n+1} \cdot f_{2n+1}^{2n+2}$. For $k < 2n - 1$, define f_k^{2n+1} by $f_k^{2n+1} = f_{2n-1}^{2n+1} \circ f_k^{2n-1}$. Then $f_1^{2n+1} : \mathfrak{A}_1 \xrightarrow{s, \equiv} \mathfrak{A}_{2n+1}$, so $\mathfrak{A}_{2n+1}$ is a model of W.

One defines $\mathfrak{A}_{2n+2}$ and $f_{2n+1}^{2n+2} : \mathfrak{A}_{2n+1} \xrightarrow{s, \Gamma} \mathfrak{A}_{2n+2} \models \mathrm{T}$ just as $\mathfrak{A}_2$ and f_1^2 were defined. For $k < 2n + 1$, define $f_k^{2n+2} = f_{2n+1}^{2n+2} \circ f_k^{2n+1}$. This defines the direct system as desired.

Now if $\mathfrak{A}_\infty = \varinjlim \mathfrak{A}_n$, then $\mathfrak{A}_\infty = \varinjlim \mathfrak{A}_{2n}$, and since T is inductive, $\mathfrak{A}_\infty$ is a model of T. But $\mathfrak{A}_\infty = \varinjlim \mathfrak{A}_{2n+1}$, and so by Theorem 8.5, $f_1^\infty : \mathfrak{A}_1 \xrightarrow{\equiv} \mathfrak{A}_\infty$. Thus $\mathfrak{A}_1 = \mathfrak{A}$ is a model of T, and since S possesses a strongly characteristic set Γ_{S}, it follows that T and W are equivalent, as desired. ■

COROLLARY 10.7. Under the hypothesis of 10.6, a formula $\mathbf{A} \in \mathrm{ML}$ is inductive iff it is equivalent in T to a $\Box\forall$-$\Diamond\exists$-formula.

Proof. Proceed as for Corollary 9.8.

§11. JOINT CONSISTENCY AND INTERPOLATION

If $\mathfrak{A}$ is a modal structure for the language ML and if ML′ is a sublanguage of ML, we define the *restriction of* $\mathfrak{A}$ *to* ML′, $\mathfrak{A}\restriction \text{ML}'$, as follows. Let $\mathfrak{A} = \langle \mathscr{A}_k, \mathbf{K}, \mathbf{L}, \mathbf{O}, \mathbf{N} \rangle$ and let L′ be the underlying classical language of ML′. Then $\mathfrak{A}\restriction \text{ML}' = \langle \mathscr{A}_k\restriction \text{L}', \mathbf{K}, \mathbf{L}, \mathbf{O}, \mathbf{N} \rangle$, where $\mathscr{A}_k\restriction \text{L}'$ is the restriction of the classical structure $\mathscr{A}_k$ to L′ (cf. Shoenfield (1967), p. 43). We also say that $\mathfrak{A}$ is an *expansion* of $\mathfrak{A}\restriction \text{ML}'$ to ML. Let S be a modal system and let T and T′ be S-theories with languages ML(T) and ML(T′), respectively. We define the *union of* T *and* T′, T ∪ T′, to be the S-theory with language L″ whose nonlogical symbols consist of precisely those of ML(T) together with those of ML(T′) (at this point, if we have not already, we agree to the convention of Shoenfield (1967), p. 15, that if a given symbol is used in one manner in a certain language, it is used in all other languages in the identical manner), and whose nonlogical axioms are precisely those of T together with those of T′. The following proof of the Joint Consistency Theorem is a rather direct adaptation of the classical proof as given in Shoenfield (1967). An approach using somewhat different techniques is contained in Gabbey (1972a).

JOINT CONSISTENCY THEOREM (11.1). Let S be a modal system having the weak tree property and possessing a strongly characteristic set, and let T and T′ be S-theories. Then T ∪ T′ is S-inconsistent iff there exists a closed formula **A** common to ML(T) and ML(T′) such that $\vdash^{\text{S}}_{\text{T}} \mathbf{A}$ and $\vdash^{\text{S}}_{\text{T}'} \neg \mathbf{A}$.

Proof. Clearly, if such a formula exists, then T ∪ T′ is S-inconsistent. For the converse, assume that no such formula exists. We will show that T ∪ T′ is consistent. Let Γ be the set of closed formulas **A** of ML(T) such that $\vdash^{\text{S}}_{\text{T}'} \mathbf{A}$. Then $\text{T}[\Gamma]$ is consistent. For otherwise there are axioms $\mathbf{A}_1, \ldots, \mathbf{A}_n$ of T and $\mathbf{B}_1, \ldots, \mathbf{B}_m \in \Gamma$ such that

$$\vdash^{\text{S}} \neg \mathbf{A}_1 \vee \ldots \vee \neg \mathbf{A}_n \vee \neg \mathbf{B}_1 \vee \ldots \vee \neg \mathbf{B}_m,$$

so that $\vdash^{\text{S}}_{\text{T}} \neg(\mathbf{B}_1 \wedge \ldots \wedge \mathbf{B}_m)$. But since $\vdash^{\text{S}}_{\text{T}'} \mathbf{B}_i$ for $i = 1, \ldots, m$, we have $\vdash^{\text{S}}_{\text{T}'} \mathbf{B}_1 \wedge \ldots \wedge \mathbf{B}_m$, contradicting our hypothesis.

Now let ML be the modal language whose nonlogical symbols are precisely those common to ML(T) and ML(T′). We will define a direct system $\{\mathfrak{A}_n, f_n^m\}$ of S-structures for L(T ∪ T′) such that each f_n^m is strong and:

(a) $\{\mathfrak{A}_{2n+1}, f_{2n+1}^{2m+1}\}$ is an elementary direct system of models of T;
(b) $\{\mathfrak{A}_{2n}, f_{2n}^{2m}\}$ is an elementary direct system of models of T′;
(c) $\{\mathfrak{A}_n \restriction \mathrm{ML}, f_n^m\}$ is an elementary direct system.

Let Δ consist of all formulas of ML so that Δ is regular and includes all ∗∗-formulas of ML. Since S possesses a strongly characteristic set, let $\mathfrak{A}_1$ be any S-model of $\mathrm{T}[\Gamma]$. Let $\mathfrak{B}$ be an expansion of $\mathfrak{A}_1 \restriction \mathrm{ML}$ to ML(T′). Then the same formulas of ML are valid in $\mathfrak{A}_1$ and $\mathfrak{B}$. Let **A** be a formula of ML such that $\vdash^{\mathrm{S}}_{\mathrm{T}'} \mathbf{A}$, and let A′ be the closure of **A**. Then $\mathbf{A}' \in \Gamma$, so $\mathfrak{A}_1 \models \mathrm{A}'$, hence $\mathfrak{B} \models \mathbf{A}'$, and so **A** is valid in $\mathfrak{B}$. Hence by the Model Extension Theorem there is a Δ-extension $\mathfrak{A}_2$ of $\mathfrak{B}$ via a strong f_1^2 such that $\mathfrak{A}_2$ is a model of T′.

Now let $n > 1$; we construct $\mathfrak{A}_n$ as follows. Let $\mathfrak{C}$ be an expansion of $\mathfrak{A}_{2n-1} \restriction \mathrm{ML}$ to ML(T). Then if $h = f_{2n-2}^{2n-1}$, $h: \mathfrak{A}_{2n-2} \restriction \mathrm{ML} \xrightarrow{s,\equiv} \mathfrak{A}_{2n-1} \restriction \mathrm{M} = \mathfrak{C} \restriction \mathrm{ML}$, and so h is a Δ-embedding of $\mathfrak{A}_{2n-2}$ in $\mathfrak{C}$. Thus by Theorem 9.4 there exists an $\mathfrak{A}_{2n}$ together with maps $f_{2n-1}^{2n}: \mathfrak{A}_{2n-1} \xrightarrow{s,\Delta} \mathfrak{A}_{2n}$ and $f_{2n-2}^{2n}: \mathfrak{A}_{2n-2} \xrightarrow{s,\equiv} \mathfrak{A}_{2n}$ such that $f_{2n-2}^{2n} = f_{2n-1}^{2n} \circ f_{2n-2}^{2n-1}$. For $k < 2n-2$, set $f_k^{2n} = f_{2n-2}^{2n} \circ f_k^{2n}$. The construction of $\mathfrak{A}_{2n+1}$, f_{2n-1}^{2n+1}, and f_{2n}^{2n+1} is similar. Note that:

(1°) with regard to ML(T ∪ T′), f_n^m is only a proto-morphism;
(2°) with regard to ML, f_n^m is an elementary monomorphism;
(3°) with regard to ML(T), f_{2n}^{2m} is an elementary monomorphism;
(4°) with regard to ML(T′), f_{2n+1}^{2n+1} is an elementary monomorphism.

Now let $\mathfrak{A} = \varinjlim \mathfrak{A}_{2n}$ and $\mathfrak{A}' = \varinjlim \mathfrak{A}_{2n+1}$. By Theorem 8, $\mathfrak{A}$ is an S-model of T and $\mathfrak{A}'$ is an S-model of T. Clearly $\mathfrak{A} \restriction \mathrm{ML} = \varinjlim \mathfrak{A}_n \restriction \mathrm{ML} = \mathfrak{A}' \restriction \mathrm{ML}$. From this it follows that $b(\mathfrak{A}) = b(\mathfrak{A}')$, and if $\mathfrak{A} = \langle \mathscr{A}_k, \mathbf{K}, \mathbf{R}, \mathbf{O}, \mathbf{N} \rangle$, then for each $k \in \mathbf{K}$, $\mathscr{A}_k \restriction \mathrm{L} = \mathscr{A}'_k \restriction \mathrm{L}$, where $\mathfrak{A}' = \langle \mathscr{A}'_k, \mathbf{K}, \ldots \rangle$, and where L is the classical sublanguage of ML. Then for each $k \in \mathbf{K}$ there is a classical structure $\mathfrak{B}_k$ for ML(T ∪ T′) the classical sublanguage of T ∪ T′, such that $\mathscr{B}_k \restriction \mathrm{L(T)} = \mathscr{A}_k$ and $\mathscr{B}_k \restriction \mathrm{L(T')} = \mathscr{A}'_k$. Let $\mathfrak{B} = \langle \mathscr{B}_k, \mathbf{K}, \mathbf{R}, \mathbf{O}, \mathbf{N} \rangle$. Since $\mathfrak{B} \restriction \mathrm{ML(T)} = \mathfrak{A}$ and $\mathfrak{B} \restriction \mathrm{ML(T')} = \mathfrak{A}'$, then $\mathfrak{B}$ is a model of T ∪ T′, as desired. ■

INTERPOLATION THEOREM (11.2). Let S be a modal system having the weak tree property possessing a strongly characteristic set Γ_{S}. Let T and T′ be S-theories and let **A** be a formula of ML(T) and **B** be a formula of ML(T′). If $\vdash^{\mathrm{S}}_{\mathrm{T} \cup \mathrm{T}'} \mathbf{A} \to \mathbf{B}$, then there exists a formula **C**, necessarily belonging to the common part of ML(T) and ML(T′), such that $\vdash^{S}_{T}\mathbf{A} \to \mathbf{C}$ and $\vdash^{S}_{T'}\mathbf{C} \to \mathbf{B}$.

Proof. First assume that both **A** and **B** are closed. In the S-theory $T[A] \cup T'[\neg B]$, we can clearly prove **A**, ¬**B**, **A**→**B**, and **B**, since $T \cup T'$ is a subtheory. Hence $T[A] \cup T'[\neg B]$ is inconsistent. By the Joint Consistency Theorem there exists a closed formula **C** such that $\vdash^{S}_{T[A]}\mathbf{C}$ and $\vdash^{S}_{T'[\neg B]}\neg\mathbf{C}$. Thus there are nonlogical axioms $\mathbf{D}_1, \ldots, \mathbf{D}_n$ of T and $\mathbf{E}_1, \ldots, \mathbf{E}_m$ of T′ such that

$$\vdash^{S}\neg\mathbf{D}_1 \vee \ldots \vee \neg\mathbf{D}_n \vee \neg\mathbf{A} \vee \mathbf{C} \quad \text{and}$$
$$\vdash^{S}\neg\mathbf{E}_1 \vee \ldots \vee \neg\mathbf{E}_m \vee \neg\neg\mathbf{B} \vee \neg\mathbf{C}.$$

It follows that $\vdash^{S}_{T}\mathbf{A} \to \mathbf{C}$ and $\vdash^{S}_{T'}\mathbf{C} \to \mathbf{B}$.

When **A** and **B** are not necessarily closed, let $\mathbf{x}_1, \ldots, \mathbf{x}_n$ be all the variables occurring free in either **A** or **B**, and let $\mathbf{e}_1, \ldots, \mathbf{e}_n$ be new individual constants not occurring in the language at hand. Let **A**′ be $\mathbf{A}_{\mathbf{x}_1 \ldots \mathbf{x}_n}[\mathbf{e}_1, \ldots, \mathbf{e}_n]$ and let **B**′ be $\mathbf{B}_{\mathbf{x}_1 \ldots \mathbf{x}_n}[\mathbf{e}_1, \ldots, \mathbf{e}_n]$. Then

$$\vdash^{S}_{V \cup V'}\mathbf{A}' \to \mathbf{B}',$$

where V(V′) is the result of adding $\mathbf{e}_1, \ldots, \mathbf{e}_n$ to T(T′). By the foregoing there is a closed $\mathbf{C}' \in ML(V) \cap ML(V')$ such that $\vdash_{V}\mathbf{A}' \to \mathbf{C}'$ and $\vdash_{V'}\mathbf{C}' \to \mathbf{B}'$. By changing bound variables if necessary, we can assume that none of $\mathbf{x}_1, \ldots, \mathbf{x}_n$ occur bound in **C**′. Let **C** result from **C**′ by replacing $\mathbf{e}_i$ by $\mathbf{x}_i$ for $i = 1, \ldots, n$. Then $\mathbf{C} \in ML(T) \cap ML(T')$ and by the Theorem on Constants, $\vdash^{S}_{T}\mathbf{A} \to \mathbf{C}$ and $\vdash^{S}_{T'}\mathbf{C} \to \mathbf{B}$, as desired. ■

DEFINITION 11.3. Let S be a modal system, let T be an S-theory, let *Q* be a set of nonlogical symbols of T and let **p** and **f** be *n*-ary nonlogical symbols of T which are not in *Q*. The predicate symbol **p** is *(explicitly) definable in terms of Q in* S *and* T if there is a formula **A** containing no nonlogical symbols not in *Q* such that if $\mathbf{x}_1, \ldots, \mathbf{x}_n$ are distinct variables, $\vdash^{S}_{T}\mathbf{p}\mathbf{x}_1 \ldots \mathbf{x}_n \leftrightarrow \mathbf{A}$. The function symbol **f** is (*explicitly*) *definable in terms of Q in* S *and* T if there is a formula **A** containing no nonlogical symbols not in *Q* such that if $\mathbf{x}_1, \ldots, \mathbf{x}_n, \mathbf{y}$ are distinct variables, then $\vdash^{S}_{T}\mathbf{f}\mathbf{x}_1 \ldots \mathbf{x}_n = \mathbf{y} \leftrightarrow \mathbf{A}$.

DEFINITION 11.4. Let S be a modal system, let T be an S-theory, let *Q* be a set of nonlogical symbols of T and let **p** and **f** be *n*-ary nonlogical symbols of T which are not in *Q*. Let **p**′ and **f**′ be new *n*-ary nonlogical symbols and let T′ be just like T except that **p** (resp. **f**) is replaced everywhere by **p**′ (resp. **f**′). Then **p** is *implicitly definable in terms of Q* *in* S *and* T if there is a formula $\mathbf{A(p)}$ of ML(T) containing no nonlogical symbols not in $Q \cup \{\mathbf{p}\}$ such that if $\mathbf{x}_1, \ldots, \mathbf{x}_n$ are pairwise distinct, then $\vdash^S_T \mathbf{A(p)}$ and

$$\vdash^S_{T \cup T'} \mathbf{A(p)} \wedge \mathbf{A(p')} \rightarrow [\mathbf{px}_1 \ldots \mathbf{x}_n \rightarrow \mathbf{p'x}_1 \ldots \mathbf{x}_n],$$

where $\mathbf{A(p')}$ is the formula of ML(T′) obtained from $\mathbf{A(p)}$ by replacing each occurrence of $\mathbf{p}$ by $\mathbf{p}'$. We say that $\mathbf{f}$ is *implicitly definable in terms of Q in* S *and* T if there is a formula $\mathbf{A(f)}$ of ML(T) containing no nonlogical symbols not in $Q \cup \{\mathbf{f}\}$ such that if $\mathbf{x}_1, \ldots, \mathbf{x}_n, \mathbf{y}$ are pairwise distinct, then $\vdash^S_T \mathbf{A(f)}$ and

$$\vdash^S_{T \cup T'} \mathbf{A(f)} \wedge \mathbf{A(f')} \rightarrow [\mathbf{fx}_1 \ldots \mathbf{x}_n = \mathbf{y} \rightarrow \mathbf{f'x}_1 \ldots \mathbf{x}_n = \mathbf{y}],$$

where $\mathbf{A(f')}$ is the formula of ML(T′) obtained from $\mathbf{A(f)}$ by replacing each occurrence of $\mathbf{f}$ by $\mathbf{f}'$.

DEFINABILITY THEOREM (11.5). Let S be a modal system having the weak tree property possessing a strongly characteristic set Γ_S and let T be an S-theory. Let Q be a set of nonlogical symbols of ML(T) and let $\mathbf{u}$ be a nonlogical symbol of T not in Q. Then $\mathbf{u}$ is definable in terms of Q in S and T iff u is implicitly definable in terms of Q in S and T.

Proof. We will treat the case in which $\mathbf{u}$ is an n-ary predicate symbol $\mathbf{p}$. Suppose that $\mathbf{p}$ is definable in terms of Q in S and T, say $\vdash^S_T \mathbf{px}_1 \ldots \mathbf{x}_n \leftrightarrow \mathbf{B}$, where $\mathbf{B}$ contains no nonlogical symbols not in Q. Then the desired $\mathbf{A}(p)$ is just the formula $\mathbf{px}_1 \ldots \mathbf{x}_n \leftrightarrow \mathbf{B}$. Conversely, suppose that $\mathbf{p}$ is implicitly definable in terms of Q in S and T, say via $\mathbf{A(p)}$ where $\mathbf{A(p)}$ contains no nonlogical symbols not in $Q \cup \{\mathbf{p}\}$. We have $\vdash^S_{T \cup T'} \mathbf{A(p)} \wedge \mathbf{A(p')} \rightarrow [\mathbf{px}_1 \ldots \mathbf{x}_n \rightarrow \mathbf{p'x}_1 \ldots \mathbf{x}_n]$. Thus we have $\vdash^S_{T \cup T'} \mathbf{px}_1 \ldots \mathbf{x}_n \rightarrow \mathbf{p'x}_1 \ldots \mathbf{x}_n$, since $\vdash^S_T \mathbf{A(p)}$ and this clearly implies that $\vdash^S_{T'} \mathbf{A(p')}$. Hence by the Interpolation Theorem, there exists a formula $\mathbf{B}$ all of whose nonlogical symbols are common to T and T′ (in particular, none can be $\mathbf{p}$ or $\mathbf{p}'$) such that $\vdash^S_T \mathbf{px}_1 \ldots \mathbf{x}_n \rightarrow \mathbf{B}$ and $\vdash^S_{T'} \mathbf{B} \rightarrow \mathbf{p'x}_1 \ldots \mathbf{x}_n$. Let $\mathfrak{P}'$ be the proof of $\mathbf{B} \rightarrow \mathbf{p'x}_1 \ldots \mathbf{x}_n$ in S from T′. It is easy to see that if $\mathfrak{P}$ is the result of replacing $\mathbf{p}'$ everywhere in $\mathfrak{P}'$ by $\mathbf{p}$, then $\mathfrak{P}$ is a proof of $\mathbf{B} \rightarrow \mathbf{px}_1 \ldots \mathbf{x}_n$ in S from T. It follows that $\vdash^S_T \mathbf{px}_1 \ldots \mathbf{x}_n \leftrightarrow \mathbf{B}$, as desired. ■

§12. MODEL COMPLETENESS

We say that an S-theory T is *model complete* if for every pair $\mathfrak{A}$, $\mathfrak{B}$ of S-models of T and every m, if $m: \mathfrak{A} \to \mathfrak{B}$, then $m: \mathfrak{A} \overset{=}{\to} \mathfrak{B}$.

LEMMA 12.1. Let S possess a strongly characteristic set Γ_S and let T be a model complete S-theory. Then for each S-model $\mathfrak{A}$ of T which is a weak tree, $T \cup MD(\mathfrak{A})$ is an S-complete S-theory.

Proof. Let $\mathfrak{A}$ be an S-model of T and let $\mathfrak{B}_1$ and $\mathfrak{B}_2$ be S-models of $T \cup MD(\mathfrak{A})$. By Theorem 6.11 there exist ultrafilter pairs $\langle I_1, F_1 \rangle$ and $\langle I_2, F_2 \rangle$ and maps m_1, m_2 such that $m_i: \mathfrak{A} \to (\mathfrak{B}_i)^{I_i}_{F_i}$ for $i = 1, 2$. For $i = 1, 2$, let $\mathfrak{C}_i$ be $(\mathfrak{B}_i)^{I_i}_{F_i}$ and let $d_i: \mathfrak{B}_i \to \mathfrak{C}_i$ be the canonical embedding. By Theorem 6.5 each of d_1 and d_2 is an elementary embedding; hence both $\mathfrak{C}_1$ and $\mathfrak{C}_2$ are models of T. Since T is model complete, then both m_1 and m_2 are elementary embeddings. Let **A** be a sentence of the underlying modal language ML. Then since all the maps in sight are elementary, we have

$$\mathfrak{B}_1 \models \mathbf{A} \quad \text{iff} \quad \mathfrak{C}_1 \models \mathbf{A} \quad \text{iff} \quad \mathfrak{A} \models \mathbf{A} \quad \text{iff} \quad \mathfrak{C}_2 \models \mathbf{A} \quad \text{iff} \quad \mathfrak{B}_2 \models \mathbf{A}.$$

Hence $\mathfrak{B}_1 \equiv \mathfrak{B}_2$ and so, since S possesses a strongly characteristic set Γ_S, it follows that $T \cup MD(\mathfrak{A})$ is S-complete. ■

The next theorem extends results of Robinson to our modal setting (cf. Sacks (1972), Section 8).

THEOREM 12.2. Let S have the weak tree property and possess a strongly characteristic set Γ_S and let T be an S-theory. The following are equivalent.

(i) T is model complete.
(ii) For each S-model $\mathfrak{A}$ of T, $T \cup MD(\mathfrak{A})$ is S-complete.
(iii) For each formula **A** of ML(T) there is a $\Diamond\exists$-formula **B** such that $\vdash^S_T \mathbf{A} \leftrightarrow \mathbf{B}$.

The implication (i)⇒(ii) is just Lemma 12.1. Next assume (iii) and let $m: \mathfrak{A} \to \mathfrak{B}$ where $\mathfrak{A}$ and $\mathfrak{B}$ are S-models of T. Let **A** be a formula in ML(T). By (iii) there is a $\Diamond\exists$-formula **B** such that $\vdash^S_T \mathbf{A} \leftrightarrow \mathbf{B}$. Let v be an assignment in $\mathfrak{A}$ and suppose that $\mathfrak{A} \models \mathbf{A}[v]$. Since $\mathfrak{A}$ is an S-model of T, then $\mathfrak{A} \models \mathbf{B}[v]$ and so by Lemma 10.1, $\mathfrak{B} \models \mathbf{B}[m \circ v]$. Then since $\mathfrak{B}$ is an S-model of T, $\mathfrak{B} \models \mathbf{A}[m \circ v]$ and so m is a weak elementary embedding.

Finally, assume (ii) and let $\mathbf{A}$ be a formula of ML(T) whose free variables are among $\mathbf{x}_1, \ldots, \mathbf{x}_n$. Let Q be the set of formulas $\mathbf{B}$ with free variables among $\mathbf{x}_1, \ldots, \mathbf{x}_n$ such that $\mathbf{B}$ is a $\Diamond\exists$-formula and $\vdash^{\mathrm{S}}_{\mathrm{T}} \mathbf{B} \to \mathbf{A}$. Let $\mathbf{e}_1, \ldots, \mathbf{e}_n$ be new constants and let U be obtained from T by adding the following axioms:

(1°) $\mathbf{A}_{\mathbf{x}_1 \ldots \mathbf{x}_n}[\mathbf{e}_1, \ldots, \mathbf{e}_n]$, and

(2°) $\neg \mathbf{B}_{\mathbf{x}_1 \ldots \mathbf{x}_n}[\mathbf{e}_1, \ldots, \mathbf{e}_n]$, where $\mathbf{B}$ belongs to Q.

Suppose U were consistent. Then since S has the weak tree property and possesses a strongly characteristic set Γ_{S}, U would have an S-model $\mathfrak{A}$ which is a weak tree. Since $\mathfrak{A}$ is an S-model of $\mathrm{T} \cup \mathrm{MD}(\mathfrak{A})$ and $\mathfrak{A}$ is a model of (1°) above, it follows that

$$\vdash^{\mathrm{S}}_{\mathrm{T} \cup \mathrm{MD}(\mathfrak{A})} \mathbf{A}_{\mathbf{x}_1 \ldots \mathbf{x}_n}[\mathbf{e}_1, \ldots, \mathbf{e}_n].$$

It follows that there are $\mathbf{C}_1, \ldots, \mathbf{C}_s$ in MD($\mathfrak{A}$) such that

$$\vdash^{\mathrm{S}}_{\mathrm{T}} \mathbf{C}_1 \wedge \ldots \wedge \mathbf{C}_s \to \mathbf{A}_{\mathbf{x}_1 \ldots \mathbf{x}_n}[\mathbf{e}_1, \ldots, \mathbf{e}_n].$$

Let $\mathbf{D}$ be $\mathbf{C}_1 \wedge \ldots \wedge \mathbf{C}_s$. Then since each $\mathbf{C}_i$ is a $\Diamond\exists$-formula, $\mathbf{D}$ is a $\Diamond\exists$-formula and so $\mathbf{D}$ belongs to Q. Thus by the construction of U, $\vdash^{\mathrm{S}}_{\mathrm{U}} \neg \mathbf{D}$ and so $\mathfrak{A} \models \neg \mathbf{D}$ since $\mathfrak{A}$ is a model of U. On the other hand, for $i = 1, \ldots, s$, $\mathbf{C}_i$ belongs to MD($\mathfrak{A}$) and so $\mathfrak{A} \models \mathbf{C}_i$. Hence $\mathfrak{A} \models \mathbf{D}$, a contradiction. Thus U must be inconsistent. Hence there are $\mathbf{B}_1, \ldots, \mathbf{B}_p$ in Q such that

$$\vdash^{\mathrm{S}}_{\mathrm{T}} [\neg \mathbf{A} \vee \mathbf{B}_1 \vee \ldots \vee \mathbf{B}_p]_{\mathbf{x}_1, \ldots, \mathbf{x}_n}[\mathbf{e}_1, \ldots, \mathbf{e}_n].$$

Then by the Theorem on Constants (1.6), $\vdash^{\mathrm{S}}_{\mathrm{T}} \neg \mathbf{A} \vee \mathbf{B}_1 \vee \ldots \vee \mathbf{B}_p$, so that

$$\vdash^{\mathrm{S}}_{\mathrm{T}} \mathbf{A} \to \mathbf{B}_1 \vee \ldots \vee \mathbf{B}_p.$$

But each $\mathbf{B}_i$ belongs to Q for $i = 1, \ldots, p$, and so $\vdash^{\mathrm{S}}_{\mathrm{T}} \mathbf{B}_i \to \mathbf{A}$. Hence

$$\vdash^{\mathrm{S}}_{\mathrm{T}} \mathbf{B}_1 \vee \ldots \vee \mathbf{B}_p \to \mathbf{A}.$$

Since each $\mathbf{B}_i$ is a $\Diamond\exists$-formula, so is $\mathbf{B}_1 \vee \ldots \vee \mathbf{B}_p$, completing the proof. ■

From Theorem 8.5 it is clear that any model complete S-theory T is inductive. Then Theorem 10.7 yields the following:

THEOREM 12.3. Let S possess a strongly characteristic set Γ_S and let T be a model complete S-theory. Then T is equivalent in S to a theory all of whose axioms are $\Box\forall$-$\Diamond\exists$-formulas.

The next theorem provides yet another condition equivalent to model completeness; we will find it useful later.

THEOREM 12.4. Let S possess a strongly characteristic Γ_S and let T be an S-theory. Then T is model complete iff for S-models $\mathfrak{A}$ and $\mathfrak{B}$ of T which are weak trees and all $f: \mathfrak{A} \to \mathfrak{B}$, there exist $\mathfrak{C}$, g, and h such that Figure 5 is commutative.

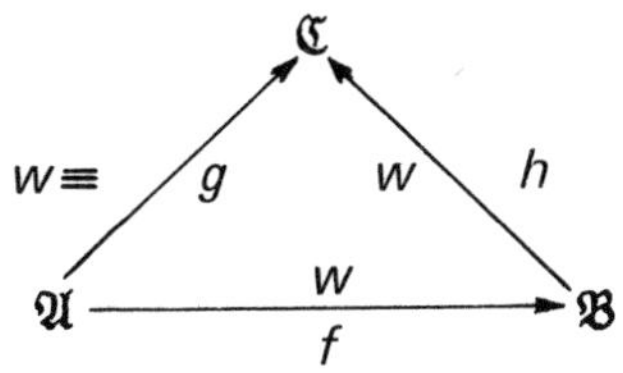

Fig. 5.

Proof. If T is model complete, $\mathfrak{A}$ and $\mathfrak{B}$ are models of T, and $f: \mathfrak{A} \to \mathfrak{B}$, then f is elementary and so it suffices to take $\mathfrak{C} = \mathfrak{B}$ and $g = h = f$. Conversely, assume the given condition, let $\mathfrak{A}_0$ and $\mathfrak{B}_0$ be S-models of T which are weak trees, and let $f_0: \mathfrak{A}_0 \to \mathfrak{B}_0$. Then by the condition there exist $\mathfrak{A}_1$, g_{01}, and h_0 such that Figure 6 is commutative.

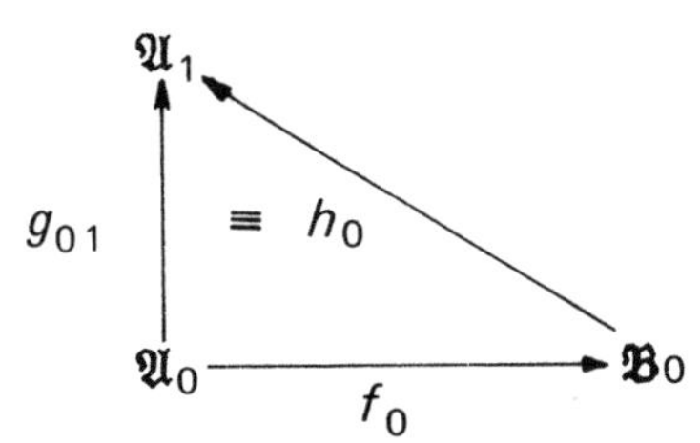

Fig. 6.

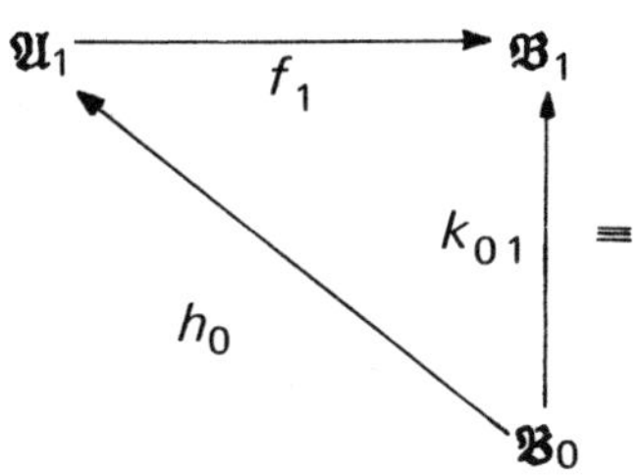

Fig. 7.

Since g_{01} is elementary, it follows that $\mathfrak{A}_1$ is an S-model of T. Consequently, again using the condition, there exist f_1, k_{01}, and $\mathfrak{B}_1$ such that Figure 7 is commutative.

Then $\mathfrak{B}_1$ is also an S-model of T since k_{01} is elementary. Iterating this procedure, we can construct the following infinitely proceeding diagram (Figure 8),

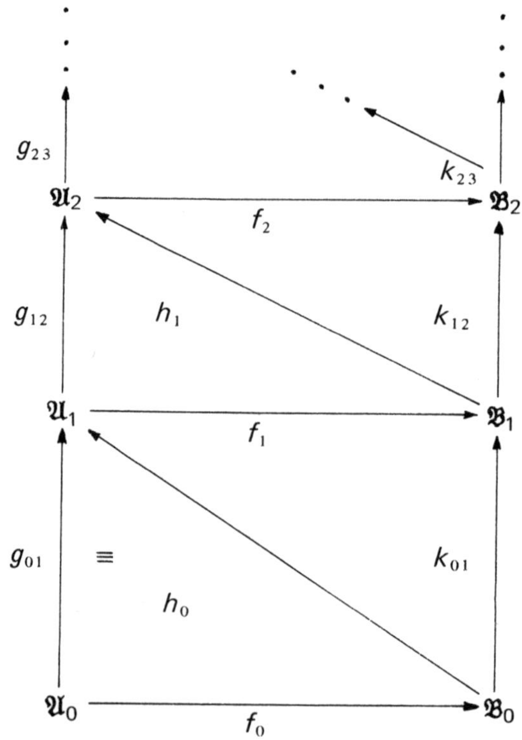

where for $i=0,1,2,\ldots$, $\mathfrak{A}_i$ and $\mathfrak{B}_i$ are S-models of T. Let $\mathfrak{A}_\infty = \varinjlim \mathfrak{A}_i$ and $\mathfrak{B}_\infty = \varinjlim \mathfrak{B}_i$. By Theorem 8.5, for all i, $g_{i\infty}: \mathfrak{A}_i \Rrightarrow \mathfrak{A}_\infty$ and $k_{i\infty}: \mathfrak{B}_i \Rrightarrow \mathfrak{B}_\infty$. Define f_∞ on $\mathfrak{A}_\infty$ as follows. Let $k \in \mathbf{K}^\infty$. Choose an i and a $k_i \in \mathbf{K}^i$ such that $g_{i\infty}(k^i) = k$. Then set $f_\infty(k) = (k_{i\infty} \circ f_i)(k_i)$. This is well-defined. For if $k_j \in \mathbf{K}^j$ is such that $g_{j\infty}(k_j) = k$, then $g_{i\infty}(k_i) = g_{j\infty}(k_j)$ implies that there is an $l \geq i, j$ such that $g_{il}(k_i) = g_{jl}(k_j)$, call this k_l. Then:

$$(k_{i\infty} \circ f_i)(k_i) = (k_{l\infty} \circ k_{il} \circ f_i)(k_i) = (k_{l\infty} \circ f_l \circ g_{il})(k_i) = (k_{l\infty} \circ f_l)(k_l).$$

and similarly, $(k_{j\infty} \circ f_j)(k_j) = (k_{l\infty} \circ f_l)(k_l)$. Now let $a \in |\mathscr{A}_k^\infty|$, and choose i so that there are $k_i \in \mathbf{K}^i$ and $a_i \in |\mathscr{A}_{k_i}^i|$ such that $g_{i\infty}(k_i) = k$ and $g_{i\infty}(a_i) = a$. Then define $f_\infty(a) = (k_{i\infty} \circ f_i)(a_i)$. As above, this is easily seen to be well-defined.

Now suppose that $k\mathbf{R}^\infty k'$. Then there is an i such that there are $k_i, k_i' \in \mathbf{K}^i$ such that $g_{i\infty}(k_i) = k$, $g_{i\infty}(k_i') = k'$, and $k_i R^i k_i'$. Then $f_i(k_i)\mathbf{S}^i f_i(k_i')$, so $(k_{i\infty} \circ f_i)(k_i)\mathbf{S}^\infty (k_{i\infty} \circ f_i)(k_i')$, and hence $f_\infty(k)\mathbf{S}^\infty f_\infty(k')$. Now suppose that

$a, a' \in |\mathscr{A}_k^\infty|$ and $\neg a \equiv_k^\infty a'$. Let i be such that there is a $k_i \in \mathbf{K}^i$ and $a_i, a_i' \in |\mathscr{A}_{k_i}|$ such that $g_{i\infty}(k_i) = k$, $g_{i\infty}(a_i) = a$, and $g_{i\infty}(a_i') = a'$. Then $\neg a_i \equiv_{k_i} a_i'$. Hence, since f_i is a monomorphism, $\neg f_i(a_i) \equiv_{f_i(k_i)} f_i(a_i')$ and since $k_{i\infty}$ is a monomorphism, $\neg f_\infty(a) \equiv_{f_\infty(k)}^\infty f_\infty(a')$. In a similar vein it is easy to compute that f_∞ is an elementary embedding of $\mathfrak{A}_\infty$ in $\mathfrak{B}_\infty$. Moreover, f_∞ is onto $\mathrm{sk}(\mathfrak{B}_\infty)$. For let $z \in \mathrm{sk}(\mathfrak{B}_\infty)$. Then for some i there is a $z_i \in \mathrm{sk}(\mathfrak{B}_i)$ such that $k_{i\infty}(z_i) = z$. But then $h_i(z_i) \in \mathrm{sk}(\mathfrak{A}_{i+1})$ and

$$\begin{aligned} f_\infty((g_{i+1\infty} \circ h_i)(z_i)) &= (k_{i+1\infty} \circ f_{i+1} \circ h_i)(z_i) = (k_{i+1\infty} \circ k_{ii+1})(z_i) \\ &= k_{i\infty}(z_i) = z. \end{aligned}$$

Finally, let $\mathbf{A}$ be a formula and let v be an assignment in $\mathfrak{A}_0$. Then:

$$\mathfrak{A}_0 \models \mathbf{A}[v] \quad \text{iff} \quad \mathfrak{A}_\infty \models \mathbf{A}[g_{0\infty} \circ v] \quad \text{iff} \quad \mathfrak{B}_\infty \models \mathbf{A}[f_\infty \circ g_\infty \circ v].$$

But for any $z \in \mathrm{sk}(\mathfrak{A}_0)$, $(f_\infty \circ g_{0\infty})(z) = (k_{0\infty} \circ f_0)(z)$. Hence,

$$\begin{aligned} &\mathfrak{B}_\infty \models \mathbf{A}[f_\infty \circ g_{0\infty} \circ v] \quad \text{iff} \quad \mathfrak{B}_\infty \models \mathbf{A}[k_{0\infty} \circ f_0 \circ v] \quad \text{iff} \\ &\mathfrak{B}_0 \models \mathrm{A}[f_0 \circ v]. \end{aligned}$$

Thus f_0 is an elementary embedding, as desired. ■

DEFINITION 12.5. Let T and T′ be S-theories with the same language. Then T′ is a *model-completion* of T if all of the following hold:

(i) T′ is an extension of T (i.e., $\vdash_{\mathrm{T}}^{\mathrm{S}} \mathbf{A}$ implies $\vdash_{\mathrm{T}'}^{\mathrm{S}} \mathbf{A}$).

(ii) If $\mathfrak{A}$ is a weak tree which is an S-model of T, there exist $\mathfrak{B}$ and f such that $f: \mathfrak{A} \to \mathfrak{B}$ is an S-model of T′.

(iii) If $\mathfrak{A}$ is a model of T and $\mathfrak{B}_1$ and $\mathfrak{B}_2$ are both models of T′ and f_1 and f_2 are such that $f_i: \mathfrak{A} \to \mathfrak{B}_i$ for $i = 1, 2$, then $(\mathfrak{B}_1, f_1(a))_{a \in |\mathscr{A}_0|} \equiv (\mathfrak{B}_2, f_2, (a))_{a \in |\mathscr{A}_0|}$.

We say that an S-model $\mathfrak{A}$ of an S-theory T *completes* T if whenever $\mathfrak{B}$ is an S-model of T and $f: \mathfrak{A} \to \mathfrak{B}$, then f is an elementary embedding. Clearly T is model complete iff every S-model of T completes T. Moreover, if T′ is a model completion of T and $\mathfrak{A}$ completes T, then $\mathfrak{A}$ is an S-model of T′. For $\mathfrak{A}$ is an S-model of T and T′ is a model completion of T, so there are $\mathfrak{B}$ and f such that $\mathfrak{B}$ is an S-model of T′ and $f: \mathfrak{A} \to \mathfrak{B}$. But then since $\mathfrak{A}$ completes T, f must be a weak elementary embedding and so $\mathfrak{A}$ is an S-model of T′.

LEMMA 12.6. Let S possess a strongly characteristic set and let T and T′ be S-theories such that T′ is a model completion of T. Then T′ is model complete.

Proof. Let $\mathfrak{A}$ be an arbitrary S-model of T′ and let $\mathfrak{B}_1$ and $\mathfrak{B}_2$ be S-models of $T' \cup MD(\mathfrak{A})$. Then by Theorem 6.13 there are ultrafilter pairs $\langle I_i, F_i \rangle$, $i = 1, 2$, such that if $\mathfrak{C}_i = (\mathfrak{B}_i)^{I_i}_{F_i}$, $i = 1, 2$, there are f_i such that $f_i : \mathfrak{A} \to \mathfrak{C}_i$. Since each $\mathfrak{B}_i$ is a model of T′ and since each canonical embedding $d_i : \mathfrak{B}_i \to \mathfrak{C}_i$ is elementary, it follows that each of $\mathfrak{C}_1$ and $\mathfrak{C}_2$ is an S-model of T′. Hence by (iii) of the definition of model completion,

$$(\mathfrak{C}_1, f_1(a))_{a \in |\mathscr{A}_\mathbf{O}|} \equiv (\mathfrak{C}_2, f_2(a))_{a \in |\mathscr{A}_\mathbf{O}|},$$

and in particular, $\mathfrak{C}_1 \equiv \mathfrak{C}_2$. Then again using the fact that each d_i is elementary, $\mathfrak{B}_1 \equiv \mathfrak{B}_2$. Hence $T' \cup MD(\mathfrak{A})$ is complete in S since S possesses a strongly characteristic set. ■

Completeness Proviso. Virtually all the theorems we wish to prove below depend on the assumption that the modal system S has the weak tree property and possesses a strongly characteristic set Γ_S. To avoid continually including this as a hypothesis, we will assume henceforth that every S considered possesses a strongly characteristic Γ_S and has the weak tree property.

In the definition introducing the notion of model completion, the wording suggests that more than one model completion of a theory might exist. However, we show next that model completions, when they exist, are unique.

THEOREM 12.7. Let T, T′, and T″ be S-theories such that both T′ and T″ are model completions of T. Then T′ and T″ are equivalent in the sense that for any $\mathbf{A}$, $\vdash^S_{T'} \mathbf{A}$ iff $\vdash^S_{T''} \mathbf{A}$.

Proof. Obviously we may assume that T, and hence also T′ and T″, are consistent. Let $\mathfrak{A}$ be an S-model of T′. We define a direct system of S-structures $\{\mathfrak{A}_n : n < \omega\}$ and morphisms $f^m_n : \mathfrak{A}_n \to \mathfrak{A}_m$, $n \leq m$, as follows. Let $\mathfrak{A}_0$ be just $\mathfrak{A}$. Now assume that $\mathfrak{A}_{2n}$ has been constructed and is a model of T′. Then $\mathfrak{A}_{2n}$ is also a model of T and so, since T″ is a model completion of T, there exists an S-structure $\mathfrak{A}_{2n+1}$ and a monomorphism $f^{2n+1}_{2n} : \mathfrak{A}_{2n} \to \mathfrak{A}_{2n+1}$ such that $\mathfrak{A}_{2n+1}$ is an S-model of T″. Finally, assume that $\mathfrak{A}_{2n+1}$ has been constructed and is an S-model of T″. Then $\mathfrak{A}_{2n+1}$ is also a model of T and again since T′ is a model completion of T, there must be an S-structure $\mathfrak{A}_{2n+2}$ and an $f^{2n+2}_{2n+1} : \mathfrak{A}_{2n+1} \to \mathfrak{A}_{2n+2}$ where $\mathfrak{A}_{2n+2}$ is an S-model of T′. Then for $n < m$, we define f^m_n by composition of the intervening maps.

$$f^m_n = f^{n+1}_n \circ f^{n+2}_{n+1} \circ \ldots \circ f^m_{m-1}.$$

Now let $\mathfrak{B} = \varinjlim \mathfrak{A}_n$. Now consider any n and the map $f_{2n}^{2n+2} : \mathfrak{A}_{2n} \to \mathfrak{A}_{2n+2}$. Both $\mathfrak{A}_{2n}$ and $\mathfrak{A}_{2n+2}$ are models of T′, and by Lemma 12.6, T′ is model complete. Thus f_{2n}^{2n+2} is elementary. Similarly, $f_{2n+1}^{2n+3} : \mathfrak{A}_{2n+1} \to \mathfrak{A}_{2n+3}$ is elementary for each n. Now it is simple to verify that $\mathfrak{B} = \varinjlim \mathfrak{A}_{2n}$ and $\mathfrak{B} = \varinjlim \mathfrak{A}_{2n+1}$. Thus by Theorem 8.5, $\mathfrak{B}$ is an elementary extension of both $\mathfrak{A}_0$ and $\mathfrak{A}_1$. Since $\mathfrak{A}_1$ is a model of T″, so is $\mathfrak{B}$ and hence $\mathfrak{A}_0 = \mathfrak{A}$ is also an S-model of T″. Thus every S-model of T′ is also an S-model of T″, and hence it follows that if $\vdash^{S}_{T''} \mathbf{A}$, then $\mathbf{A}$ is valid in every S-model of T′, and so $\vdash^{S}_{T'} A$. That $\vdash^{S}_{T'} \mathbf{A}$ implies $\vdash^{S}_{T''} \mathbf{A}$ is shown similarly. ■

§13. FINITE FORCING

Here we extend the ideas of Robinson (1970) and Barwise and Robinson (1970). Our approach is that of Keisler (1973).

Given a modal language ML, the infinitary modal language $\mathrm{ML}_{\omega_1\omega}$ is built up from ML by allowing the formation of infinite disjunctions $\vee \Phi$ or $\vee_{\mathbf{A}\in\Phi}\mathbf{A}$ for any countable set Φ of formulas. The infinite conjunction $\wedge\Phi$ or $\wedge_{\mathbf{A}\in\Phi}\mathbf{A}$ is then an abbreviation for $\neg\vee_{\mathbf{A}\in\Phi}\neg\mathbf{A}$. Since we allow the iteration of this formation of infinite disjunctions together with the usual finitary operations of ML, it follows immediately that $\mathrm{ML}_{\omega_1\omega}$ contains infinitely many formulas. In fact, $\mathrm{ML}_{\omega_1\omega}$ has $2^{\aleph_0}$ many formulas.

Given a formula $\mathbf{A}$ of $\mathrm{ML}_{\omega_1\omega}$, we define the set of subformulas of $\mathbf{A}$ recursively as follows:

(i) If $\mathbf{A}$ is atomic, $\mathrm{sub}(\mathbf{A})=\{\mathbf{A}\}$.
(ii) $\mathrm{sub}(\mathbf{A}\vee\mathbf{B})=\mathrm{sub}(\mathbf{A})\cup\mathrm{sub}(\mathbf{B})\cup\{\mathbf{A}\vee\mathbf{B}\}$.
(iii) $\mathrm{sub}(\neg\mathbf{A})=\mathrm{sub}(\mathbf{A})\cup\{\neg\mathbf{A}\}$.
(iv) $\mathrm{sub}(\exists\mathbf{x}\mathbf{A})=\mathrm{sub}(\mathbf{A})\cup\{\exists\mathbf{x}\mathbf{A}\}$.
(v) $\mathrm{sub}(\vee_{\mathbf{A}\in\Phi}\mathbf{A})=\cup_{\mathbf{A}\in\Phi}\mathrm{sub}(\mathbf{A})\cup\{\vee_{\mathbf{A}\in\Phi}\mathbf{A}\}$.
(vi) $\mathrm{sub}(\Diamond\mathbf{A})=\mathrm{sub}(\mathbf{A})\cup\{\Diamond\mathbf{A}\}$.

It is easy to verify by induction on the structure of $\mathbf{A}$ that $\mathrm{sub}(\mathbf{A})$ is at most countable.

DEFINITION 13.1 A *fragment* of $\mathrm{ML}_{\omega_1\omega}$ is a set $\mathrm{ML}_{\mathscr{A}}$ of formulas of $\mathrm{ML}_{\omega_1\omega}$ such that:

(i) Every formula of ML belongs to $\mathrm{ML}_{\mathscr{A}}$.
(ii) $\mathrm{ML}_{\mathscr{A}}$ is closed under applications of $\neg$, $\exists\mathbf{x}$, $\Diamond$, and finite disjunction.
(iii) If $\mathbf{A}\in\mathrm{ML}_{\mathscr{A}}$ and $\mathbf{a}$ is a term, then $\mathbf{A}_{\mathbf{x}}[\mathbf{a}]\in\mathrm{ML}_{\mathscr{A}}$.
(iv) $\mathrm{ML}_{\mathscr{A}}$ is closed under subformulas; i.e., $\mathbf{A}\in\mathrm{ML}$ implies that $\mathrm{sub}(\mathbf{A})\subseteq\mathrm{ML}_{\mathscr{A}}$.

In the remainder of this section, $\mathrm{ML}_{\mathscr{A}}$ will be a fixed *countable* fragment of $\mathrm{ML}_{\omega_1\omega}$. Let U be countably infinite set of new individual constants and for $u\in U$, let $C^u=\{c_0^u, c_1^u, \ldots\}$ be such that $U\cap C^u=\emptyset$ and $u\neq u'$ implies $C^u\cap C^{u'}=\emptyset$. Let $\mathrm{ML}^u_{\mathscr{A}}$ be the set of all formulas obtained from formulas of $ML_{\mathscr{A}}$ by replacing finitely many free variables by constants $c \in C^u$. If MK^u is obtained from ML by addition of the constants in C^u, then it is easy to see that $MK^u_{\mathscr{A}}$ is the least fragment of $MK^u_{\omega_1\omega}$ which contains $ML_{\mathscr{A}}$.

Let LB[U] be the language obtained from the language LB of §2 by adding the elements of U as new individual constants. Let S be a fixed modal system with characteristic set Γ_S.

DEFINITION 13.2. A *forcing property* for LB[U] is a triple $\mathscr{P}' = \langle P, \leq, h \rangle$ such that:

(i) $\langle P, \leq \rangle$ is a partially ordered set with first element 0.

(ii) h is a function defined on P such that for each $p \in P$, $h(p)$ is a set of atomic sentences of LB[U] such that:

(a) If $p \leq q$, then $h(p) \subseteq h(q)$.

(b) For each $p \in P$ and $u \in U$, there is a $q \in P$ with $p \leq q$ such that the sentences O^*R^*u and $u = u$ both belong to $h(q)$.

(c) If $u = u'$ and $\mathbf{A}_{\mathbf{x}}[u]$ both belong to $h(p)$, then there is a $q \in P$ with $p \leq q$ such that the sentence $\mathbf{A}_{\mathbf{x}}[u']$ belongs to $h(q)$.

DEFINITION 13.3. Let $\mathscr{P}' = \langle P, \leq, h \rangle$ be a forcing property for LB[U], let **A** be a sentence of LB[U], and let $p \in P$. We defined the relation $p \Vdash^{\mathscr{P}'} \mathbf{A}$, "*p forces* **A**", by induction on the structure of **A** as follows:

(i) If **A** is atomic, $p \Vdash^{\mathscr{P}'} \mathbf{A}$ iff $\mathbf{A} \in h(p)$.

(ii) $p \Vdash^{\mathscr{P}'} \neg \mathbf{A}$ iff there is no $q \geq p$ such that $q \Vdash^{\mathscr{P}'} \mathbf{A}$.

(iii) $p \Vdash^{\mathscr{P}'} \mathbf{A} \vee \mathbf{B}$ iff $p \Vdash^{\mathscr{P}'} \mathbf{A}$ or $p \Vdash^{\mathscr{P}'} \mathbf{B}$.

(iv) $p \Vdash^{\mathscr{P}'} \exists \mathbf{x} \mathbf{A}$ iff there is a $u \in U$ such that $p \Vdash^{\mathscr{P}'} \mathbf{A}_{\mathbf{x}}[u]$.

We say that p *weakly forces* **A**, written $p \Vdash^* \mathbf{A}$, iff $p \Vdash \neg\neg \mathbf{A}$.

DEFINITION 13.4. A *proto-forcing relation for* ML *over U and* $\{C^u\}$ is a quadruple $\mathscr{P} = \langle P, \leq, h, f \rangle$ such that:

(i) The triple $\mathscr{P}' = \langle P, \leq, h \rangle$ is a forcing relation for LB[U];

(ii) f is a function defined on $P \times U$ such that for each $p \in P$ and $u \in U$, $f(p, u)$ is a set of atomic sentences of $MK^{p,u}_{\mathscr{A}} = \cup \{MK^{u'}_{\mathscr{A}} : u' = u \in h(p) \vee u' R^* u \in h(p)\}$ such that:

(a) If $p \in P$ and $u = u'$ occurs in $h(p)$, then $f(p, u) = f(p, u')$.

(b) If $u \in U$ and $p, q \in P$ and $p \leq q$, then $f(p, u) \subseteq f(q, u)$.

(c) Let $p \in P$ and $u \in U$, and let **a** be a closed term of $MK^{p,u}_{\mathscr{A}}$. Then there is a $q \in P$ with $p \leq q$ such that $\mathbf{a} = \mathbf{a}$ occurs in $f(q, u)$.

(d) If $p \in P$ and $u \in U$, and if $\mathbf{a} = \mathbf{b}$ and $\mathbf{A_x}[\mathbf{a}]$ both occur in $f(p, u)$, then there is a $q \in P$ with $p \leq q$ such that $\mathbf{A_x}[\mathbf{b}]$ occurs in $f(q, u)$.

(e) For any $p \in P$ and $u \in U$ and for any closed term $\mathbf{a}$ of $\mathrm{ML}^{p;u}_{\mathscr{A}}$, there are $c \in C^u$ and $q \in P$ with $q \geq p$ such that $\mathbf{a} = c$ occurs in $f(q, u)$.

If $\mathscr{P} = \langle P, \leq, h, f\rangle$ is a proto-forcing relation and $\mathbf{A}$ is a sentence of LB[U], we write $p\Vdash^{\mathscr{P}}\mathbf{A}$ if $p\Vdash^{\mathscr{P}'}\mathbf{A}$, where $\mathscr{P}' = \langle P, \leq, h\rangle$. In general we will suppress the superscripts $\mathscr{P}$ and $\mathscr{P}'$.

DEFINITION 13.5. Let $\mathscr{P} = \langle P, \leq, h, f\rangle$ be a proto-forcing relation over C and U. Given $p \in P$, $u \in U$, and a sentence $\mathbf{A}$ of $\cup_{u\in U} \mathrm{MK}^u_{\mathscr{A}}$, we define the relation $p\Vdash_u\mathbf{A}$, *p forces* $\mathbf{A}$ *at u*, by induction on the complexity of $\mathbf{A}$ as follows:

(i) If $\mathbf{A}$ is atomic, then $p\Vdash_u\mathbf{A}$ iff $\mathbf{A} \in f(p, u)$.
(ii) $p\Vdash_u \neg \mathbf{A}$ iff there is no $q \geq p$ such that $q\Vdash_u\mathbf{A}$.
(iii) $p\Vdash_u \vee \Phi$ iff for some $\mathbf{A} \in \Phi$, $p\Vdash_u\mathbf{A}$.
(iv) $p\Vdash_u \exists \mathbf{x}\mathbf{A}$ iff for some $c \in C^u$, $p\Vdash_u\mathbf{A_x}[c]$.
(v) $p\Vdash_u \Diamond \mathbf{A}$ iff for some $u' \in U$, $p\Vdash u\mathrm{R}^* u'$ and $p\Vdash_{u'}\mathbf{A}$.

DEFINITION 13.6. If $\mathbf{A} \in \mathrm{LB}[\mathrm{U}]$, we write $p\Vdash^*\mathbf{A}$ for $p\Vdash\neg\neg\mathbf{A}$ and say that *p weakly forces* $\mathbf{A}$. And if $\mathbf{A} \in \mathrm{MK}^{\mathrm{U}}_{\mathscr{A}} = \cup_{u\in\mathrm{U}} \mathrm{MK}^u_{\mathscr{A}}$ we write $p\Vdash^*_u\mathbf{A}$ for $p\Vdash_u\neg\neg\mathbf{A}$ and say that *p weakly forces* $\mathbf{A}$ *at u*.

LEMMA 13.7. Let $\mathbf{A}$ and $\wedge\Phi$ be sentences of $\mathrm{MK}^{\mathrm{U}}_{\mathscr{A}}$, let $u \in U$, and let $p, q \in P$ where $\mathscr{P} = \langle P, \leq, h, f\rangle$ is a proto-forcing relation; and let $\mathbf{E} \in \mathrm{LB}[\mathrm{U}]$. Then:

(1) $p \leq q$ and $p\Vdash_u\mathbf{A}$ imply $q\Vdash_u\mathbf{A}$.
(2) It is not the case that both $p\Vdash_u\mathbf{A}$ and $p\Vdash_u\neg\mathbf{A}$.
(3) $\forall p \exists g \geq p[g\Vdash_u\mathbf{A}$ or $g\Vdash_u\neg\mathbf{A}]$.
(4) $p\Vdash^*_u\mathbf{A}$ iff $\forall q \geq p \exists r \geq q[r\Vdash_u\mathbf{A}]$.
(5) If $p\Vdash_u\mathbf{A}$, then $p\Vdash^*_u\mathbf{A}$.
(6) $p\Vdash^*_u\neg\mathbf{A}$ iff $p\Vdash_u\neg\mathbf{A}$.
(7) $p\Vdash_u\forall\mathbf{x}\mathbf{A}$ iff $\forall q \geq p\, \forall c \in C^u\, \exists r \geq q[r\Vdash_u\mathbf{A_x}[c]]$.
(8) $p\Vdash_u \wedge \Phi$ iff $\forall \mathbf{A} \in \Phi\, \forall q \geq p \exists r \geq q[r\Vdash_u\mathbf{A}]$.
(9) $p\Vdash_u \Box\mathbf{A}$ iff $\forall q \geq p \forall u' \in \mathrm{U} \exists r \geq q[r\Vdash uR^*u' \Rightarrow r\Vdash_{u'}\mathbf{A}]$.
(10) $p \leq q$ and $p\Vdash\mathbf{E}$ implies $q\Vdash\mathbf{E}$.
(11) It is not the case that $p\Vdash\mathbf{E}$ and $p\Vdash\neg\mathbf{E}$.

(12) $\forall p \exists q \geq p[q \Vdash \mathbf{E} \text{ or } q \Vdash \neg \mathbf{E}]$.
(13) $p \Vdash^* \mathbf{E}$ iff $\forall q \geq p \exists r \geq q[r \Vdash \mathbf{E}]$.
(14) If $p \Vdash \mathbf{E}$, then $p \Vdash^* \mathbf{E}$.
(15) $p \Vdash^* \neg \mathbf{E}$ iff $p \Vdash \neg \mathbf{E}$.
(16) $p \Vdash \forall \mathbf{x} \mathbf{E}$ iff $\forall q \geq p \forall u' \in U \exists r \geq q[r \Vdash \mathbf{E}_{\mathbf{x}}[u]]$.
(17) $p \Vdash \mathbf{E} \wedge \mathbf{F}$ iff $p \Vdash^* \mathbf{E}$ and $p \Vdash^* \mathbf{F}$.

Proof. We first prove statements (1) and (10) simultaneously by induction on the complexities of **A** and **E**; for the atomic cases, these follow immediately from the statements 13.5(i) and 13.4(iib), and 13.3(i) and 13.2(iia). If **A** is $\vee \Phi$, suppose that $p \Vdash_u \mathbf{A}$ by virtue of $p \Vdash_u \mathbf{B}$ where $\mathbf{B} \in \Phi$. By induction, $q \Vdash_u \mathbf{B}$, and so $q \Vdash_u \mathbf{A}$. Similarly if **E** is $\mathbf{F} \vee \mathbf{G}$. If **A** is $\neg \mathbf{B}$ and $q \leq r$, then $p \leq r$, so not $-q \Vdash_u \mathbf{B}$ and so $q \Vdash_u \mathbf{A}$. Similarly if **E** is $\neg \mathbf{F}$. Finally, let **A** be $\Diamond \mathbf{B}$ and suppose that for some $u' \in U$, $p \Vdash u \mathrm{R}^* u'$ and $p \Vdash_{u'} \mathbf{B}$. Then by the induction hypothesis and (10) in the atomic case, $q \Vdash u \mathrm{R}^* u'$ and $q \Vdash_{u'} \mathbf{B}$, so $q \Vdash_u \mathbf{A}$.

For (2), if $p \Vdash_u \neg \mathbf{A}$, then for all $q \geq p$, not$-q \Vdash_u \mathbf{A}$; in particular, since $p \geq p$, not$-p \Vdash_u \mathbf{A}$. Similarly for (11).

For (3), let $p \in P$ and suppose that there is no $q \geq p$ such that $q \Vdash_u \mathbf{A}$; then for all $q \geq p$, not$-q \Vdash_u \mathbf{A}$ and consequently, $p \Vdash_u \neg \mathbf{A}$. Similarly for (12).

Both (4) and (13) follow immediately from the definitions involved, while (5) and (14) follow immediately from (1) and (10), respectively.

For (6), suppose $p \Vdash_u^* \neg \mathbf{A}$ and let $q \geq p$. Then there must be an $r \geq q$ such that $r \Vdash_u \neg \mathbf{A}$, and consequently not$-r \Vdash_u \mathbf{A}$. If $q \Vdash_u \mathbf{A}$, then by (1) we would have $r \Vdash_u \mathbf{A}$, a contradiction. Thus not$-q \Vdash_u \mathbf{A}$ and so $p \Vdash_u \neg \mathbf{A}$. The converse follows from (5). Similarly for (15).

For (7), since $\forall \mathbf{x} \mathbf{A}$ is $\neg \exists \mathbf{x} \neg \mathbf{A}$,

$$\begin{aligned} p \Vdash_u \forall \mathbf{x} \mathbf{A} \quad & \text{iff} \quad \forall q \geq p \,\blacksquare\, \text{not} - q \Vdash_u \exists \mathbf{x} \neg \mathbf{A} \\ & \text{iff} \quad \forall q \geq p \,\blacksquare\, \text{not} - \exists c \in C^u \,\blacksquare\, q \Vdash_u \neg \mathbf{A}_{\mathbf{x}}[c] \\ & \text{iff} \quad \forall q \geq p \forall c \in C^u \exists r \geq q \,\blacksquare\, r \Vdash_u \mathbf{A}_{\mathbf{x}}[c]. \end{aligned}$$

Similarly for (16), while (8) and (17) are treated in a like manner.

Finally, for (9), since $\Box \mathbf{A}$ is $\neg \Diamond \neg \mathbf{A}$,

$$\begin{aligned} p \Vdash_u \Box \mathbf{A} \quad & \text{iff} \quad \forall q \geq p \,\blacksquare\, \text{not} - \exists u' \in U \,\blacksquare\, q \Vdash u \mathrm{R}^* u' \,\&\, q \Vdash_{u'} \neg \mathbf{A} \\ & \text{iff} \quad \forall q \geq p \forall u' \in U \,\blacksquare\, q \Vdash u \mathrm{R}^* u' \Rightarrow \exists r \geq q r \Vdash_{u'} \mathbf{A}. \end{aligned}$$

Assume this latter statement, let $q \geq p$ and $u' \in U$, suppose $q \Vdash u \mathrm{R}^* u$, and let $r \geq q$ be such that $r \Vdash_{u'} \mathbf{A}$. Then by (10), $r \Vdash u \mathrm{R}^* u'$, and so the right side of (17) follows. Conversely, assume the right side of (17), let $q \geq p$,

$u' \in U$, and suppose $q \Vdash u R^* u'$. By the right side of (17), there is an $r \geq q$ such that

$$r \Vdash u R^* u' \Rightarrow r \Vdash_{u'} \mathbf{A}.$$

But by (10), since $q \Vdash u R^* u'$ and $r \geq q$, then $r \Vdash u R^* u'$ and so $r \Vdash_{u'} \mathbf{A}$, and $p \Vdash_u \mathbf{A}$ follows. ■

Throughout the following, $\mathscr{P} = \langle P, \leq, h, f \rangle$ will be a fixed proto-forcing relation.

DEFINITION 13.8. A subset $G \subseteq P$ is *$\mathscr{P}$-generic for* $p_0 \in P$ provided the following all hold:

(i) $p_0 \in G$.
(ii) $p \in G \,\&\, q \leq p \Rightarrow q \in G$.
(iii) $p, q \in G \Rightarrow$ there is an $r \in G$ with $p \leq r \,\&\, q \leq r$.
(iv) For each sentence $\mathbf{A}$ of $\mathrm{MK}^u_{\mathscr{A}}$ and each $u \in U$, there is a $p \in G$ such that $p \Vdash_u \mathbf{A}$ or $p \Vdash_u \neg \mathbf{A}$.
(v) For each sentence $\mathbf{E}$ of $\mathrm{LB}[\mathrm{U}]$, there is a $p \in G$ such that $p \Vdash \mathbf{E}$ or $p \Vdash \neg \mathbf{E}$.

GENERIC SET LEMMA (13.9). For any $p_0 \in P$, there exist subsets G of P which are $\mathscr{P}$-generic for p_0.

Proof. Let $\langle \mathbf{A}_0, u_0 \rangle, \ldots, \langle \mathbf{A}_n, u_n \rangle, \ldots$ be an enumeration of $\mathrm{MK}^{\mathrm{U}}_{\mathscr{A}} \times U$ and let $\mathbf{E}_0, \mathbf{E}_1, \ldots$ be an enumeration of $\mathrm{LB}[\mathrm{U}]$. We will inductively define a sequence $p_0, p_1, \ldots$ of elements of P as follows. The element p_0 is as given. Then for $n \geq 0$, assume that p_i has been defined for $i < 2n+1$. Then using statement (3) of Lemma 13.7, let $p_{2n+1} \geq p_{2n}$ be such that

$$p_{2n+1} \Vdash_{u_n} \mathbf{A}_n \quad \text{or} \quad p_{2n+1} \Vdash_{u_n} \neg \mathbf{A}_n.$$

And then using statement (12) of Lemma 13.7, let $p_{2n+2} \geq p_{2n+1}$ be such that

$$p_{2n+2} \Vdash \mathbf{E}_n \quad \text{or} \quad p_{2n+2} \Vdash \neg \mathbf{E}_n.$$

Then, setting $G = \{q \in P\colon \text{for some } n, q \leq p_n\}$, it is easy to see that G is $\mathscr{P}$-generic for p_0. ■

Given a generic G, we will write $G \Vdash_u \mathbf{A}$ to mean that for some $p \in G$, $p \Vdash_u \mathbf{A}$, and similarly for $G \Vdash \mathbf{E}$.

DEFINITION 13.10. An *expanded modal structure* is a modal structure $\mathfrak{A} = \langle \mathscr{A}_k, \mathbf{K}, \mathbf{R}, \mathbf{O}, \mathbf{N} \rangle$ together with maps $u \to u^{\mathfrak{A}} \in \mathbf{K}$ and $c \to c^{\mathfrak{A}, k} \in |\mathscr{A}_k|$ for $c \in C^u$ where $k = u^{\mathfrak{A}}$. Such an expanded structure is said to be *canonical* if each $k \in \mathbf{K}$ is $u^{\mathfrak{A}}$ for some $u \in U$, and for each $k \in \mathbf{K}$ and $a \in |\mathscr{A}_k|$, there exist $u \in U$ and $c \in C^u$ such that $u^{\mathfrak{A}} = k$ and $a \equiv_k c^{\mathfrak{A}, k}$.

DEFINITION 13.11. Let $p_0 \in P$. An expanded modal structure $\mathfrak{A} = \langle \mathscr{A}_k, \mathbf{K}, \mathbf{R}, \mathbf{O}, \mathbf{N} \rangle$ is said to be *$\mathscr{P}$-generic for p_0* if it is canonical and there is a G which is $\mathscr{P}$-generic for p_0 such that:

(i) for any formula $\mathbf{E}$ of LB with free variables $\mathbf{w}_1, \ldots, \mathbf{w}_n$ and any $u_1, \ldots, u_n \in U$, if $v(\mathbf{w}_i) = u_i^{\mathfrak{A}}$ for $i = 1, \ldots, n$,

$$b(\mathfrak{A}) \models \mathbf{E}[v] \quad \text{iff} \quad G \Vdash \mathbf{E}_{\mathbf{w}_1 \ldots \mathbf{w}_n}[u_1, \ldots, u_n];$$

(ii) for any formula $\mathbf{A}$ of $\mathrm{ML}_{\mathscr{A}}$ with free variables $\mathbf{x}_1, \ldots, \mathbf{x}_n$, any $u \in U$, and any $c_1, \ldots, c_n \in C^u$ if $k = u^{\mathfrak{A}}$ and $v(\mathbf{x}_i) = c_i^{\mathfrak{A}, k}$ for $i = 1, \ldots, n$, then

$$\mathfrak{A} \models_k \mathbf{A}[v] \quad \text{iff} \quad G \Vdash_u \mathbf{A}_{\mathbf{x}_1 \ldots \mathbf{x}_n}[c_1, \ldots, c_n].$$

We say that G *determines* $\mathfrak{A}$. Also, we will say that $\mathfrak{A}$ is *$\mathscr{P}$-generic* if it is $\mathscr{P}$-generic for 0.

THEOREM 13.12. Let $p_0 \in P$. Given any subset G of P which is $\mathscr{P}$-generic for p_0, there is a modal structure $\mathfrak{A}$ which is $\mathscr{P}$-generic for p_0 and determined by G.

Proof. First define an equivalence relation $\sim$ on $U \cup \{\mathrm{O}^*\}$ by: $u \sim u'$ iff $G \Vdash u = u'$; it is easy to verify that this is indeed an equivalence relation. Then set $\mathbf{K} = \{u/\sim \, : u \in U\}$, and define $\mathbf{R}$ by

$$(u/\sim)\mathbf{R}(u'/\sim) \quad \text{iff} \quad G \Vdash u\mathrm{R}^*u'.$$

Suppose that $u/\sim \; = u''/\sim$ and $u'/\sim \; = u'''/\sim$, so that $G \Vdash u = u''$ and $G \Vdash u' = u'''$. Moreover, suppose that $G \Vdash u\mathrm{R}^*u'$. Let $p_1, p_2, p_3 \in G$ be such that $p_1 \Vdash u = u''$, $p_2 \Vdash u' = u'''$, and $p_3 \Vdash u\mathrm{R}^*u'$. By Definition 13.8 (iii), there is a $q \in G$ with $p_1, p_2, p_3 \leq q$; by Lemma 13.7, $q \Vdash u = u''$, $q \Vdash u' = u'''$, and $q \Vdash u\mathrm{R}^*u'$. Now Definition 13.8(v) guarantees that there is an $r \in G$ such that

$$r \Vdash u''\mathrm{R}^*u''' \quad \text{or} \quad r \Vdash \neg\, u''\mathrm{R}^*u'''.$$

Then using 13.8(iii) again, there is an $s \in G$ with $q, r \leq s$, and so by Lemma 13.7, $s \Vdash u\mathrm{R}^*u'$, $s \Vdash u = u''$, and $s \Vdash u' = u'''$. Then by Definition 13.3(i),

$u=u''$, $u'=u'''$, $u\mathrm{R}^*u' \in h(s)$, and so by two applications of Definition 13.2(c), there is an $s' \in P$ with $s \leq s'$ such that $u''\mathrm{R}^*u''' \in h(s')$, and so $s' \Vdash u''\mathrm{R}^*u'''$. Since $r \leq s'$, by Lemma 13.7 we cannot have $r \Vdash \neg u''\mathrm{R}^*u'''$, and so $r \Vdash u''\mathrm{R}^*u'''$. Thus $G \Vdash u''\mathrm{R}^*u'''$, and so **R** is well-defined. Obviously we set $\mathbf{O} = \mathrm{O}^*/\sim$. And setting

$$\mathbf{N} = \{u/\sim : G \Vdash \mathrm{N}^*(u)\},$$

we see that **N** is well-defined by an argument similar to that for **R**.

Next we must verify that:

(#) for any $\mathbf{A} \in \mathrm{MK}^{\mathrm{U}}_{\mathscr{A}}$ and any $u, u' \in U$, if $u \sim u'$ and $G \Vdash_u \mathbf{A}$, then $G \Vdash_{u'} \mathbf{A}$.

We will prove (#) by induction on the length of **A**. Let $p \in G$ be such that $p \Vdash u = u'$; then $u = u' \in h(p)$. Moreover, let $q \in G$ be such that $q \Vdash_u \mathbf{A}$, and let $r \in G$ be such that $p, q \leq r$. Then $u = u' \in h(r)$ and $r \Vdash_u \mathbf{A}$. Now if **A** is atomic, then $\mathbf{A} \in f(r, u)$; but by Definition 13.4(iia), $f(r, u) = f(r, u')$, so $r \Vdash_{u'} \mathbf{A}$ and hence $G \Vdash_{u'} \mathbf{A}$. Next suppose that **A** is $\neg \mathbf{B}$. Since $q \Vdash_u \neg \mathbf{B}$, in light of 13.8(iii) and statements (1) and (2) of Lemma 13.7, not $-q' \Vdash_u \mathbf{B}$ for all $q' \in G$. Now by the induction hypothesis with the roles of u and u' interchanged, $G \Vdash_{u'} \mathbf{B}$ implies $G \Vdash_u \mathbf{B}$, so we have not $-G \Vdash_{u'} B$. Then by Definition 13.8(iv), $G \Vdash_{u'} \neg \mathbf{B}$. For the next case, let A be $\vee \Phi$. Then for some $\mathbf{B} \in \Phi$, $q \Vdash_u \mathbf{B}$, so $G \Vdash_u \mathbf{B}$. Then by induction, $G \Vdash_{u'} \mathbf{B}$, and thus $G \Vdash_{u'} \mathbf{A}$. Finally, suppose that **A** is $\diamond \mathbf{B}$. Then there must be a $u'' \in U$ such that $q \Vdash u\mathrm{R}^*u''$ and $q \Vdash_{u''} \mathbf{B}$. Since $u = u'$ and $u\mathrm{R}^*u''$ both belong to $h(r)$, there is an $s \in P$ with $r \leq s$ and $u'\mathrm{R}^*u'' \in h(s)$. Thus by Lemma 13.7, $s \Vdash u'\mathrm{R}^*u''$ and $s \Vdash_{u''} \mathbf{B}$, so that $s \Vdash_{u'} \diamond \mathbf{B}$. Thus we have shown that if $q \in G$ and $q \Vdash_u \diamond \mathbf{B}$, there is an $s \in P$ with $q \leq s$ and $s \Vdash_{u'} \diamond \mathbf{B}$. By Definition 13.8, there is a $p' \in G$ such that either $p' \Vdash_{u'} \diamond \mathbf{B}$ or $p' \Vdash_{u'} \neg \diamond \mathbf{B}$. If the latter case were to hold, using 13.8(iii) there would be a $q' \geq q, p'$ with $q' \in G$. Then by Lemma 13.7,

$$q' \Vdash_u \diamond \mathbf{B} \quad \text{and} \quad q' \Vdash_{u'} \neg \diamond \mathbf{B}.$$

But by the remarks above, there must be an $s \geq q'$ with $s \Vdash_{u'} \diamond \mathbf{B}$, a contradiction. Thus we must have $p' \Vdash_{u'} \diamond \mathbf{B}$, and so $G \Vdash_{u'} \diamond \mathbf{B}$, which verifies (#).

Now for $k \in \mathbf{K}$, set

$$D^k = \cup \{C^u : (u/\sim) = k \quad \text{or} \quad (u/\sim)\mathbf{R}k\},$$

and let $|\mathscr{A}_k|$ consist of all closed terms of $L[D^k]$. Note that if $k \in \mathbf{K}$ here, $k \subseteq U$, so we may write $u \in k$, etc. Define $\equiv_k$ on $|\mathscr{A}_k|$ by:

COROLLARY 13.14. Let $p \in P$ and let $\mathbf{A}$ be a sentence of $ML_{\mathscr{A}}$. Then $p \Vdash_u^* \mathbf{A}$ iff for every $\mathscr{P}$-generic $\mathfrak{A}$ for p, if $k = u^{\mathfrak{A}}$, then $\mathfrak{A} \models_k \mathbf{A}$.

Proof. Similar to that of the preceding Corollary . ■

DEFINITION 13.15. Let S be a modal system which possesses a characteristic set Γ. A proto-forcing relation $\mathscr{P} = \langle P, \leq, h, f \rangle$ is said to be an S-*forcing relation* if for every sentence $\mathbf{E} \in \Gamma$, $\mathbf{O} \Vdash^* \mathbf{E}$.

COROLLARY 13.16. Let S be a modal system possessing a characteristic set Γ, and let $\mathscr{P}$ be an S-forcing relation. Then every $\mathscr{P}$-generic structure $\mathfrak{A}$ is an S-modal structure.

Proof. Immediate by Corollary 13.13. ■

§14. FORCING AND MODEL COMPLETIONS

Let L be a language of fixed similarity type and let $\mathscr{M}$ be a fixed class of S-modal structures for ML. Let Φ be a fixed set of formulas of $\mathrm{ML}_{\mathscr{A}}$ which contains all atomic formulas and is closed under subformulas. The S-forcing property $\mathscr{P}(\mathscr{M}, \Phi)$ is described as follows. Let Φ (C) be the set of all sentences of $\mathrm{MK}_{\mathscr{A}}$ of the form $\mathbf{A}_{\mathbf{x}_1 \dots \mathbf{x}_n}[c_1, \dots, c_n]$ where $\mathbf{A} \in \Phi$ and $c_1, \dots, c_n \in C = \cup_{u \in U} C$. The set P of conditions consists of all pairs $\langle \alpha, p \rangle$ meeting the following:

(i) α is a finite set of atomic sentences of LB[U];

(ii) p is a finite subset of $\Phi(C) \times U$;

(iii) there exist a structure $\mathfrak{A} = \langle \mathscr{A}_k, \mathbf{K}, \mathbf{R}, \mathbf{O}, \mathbf{N} \rangle$ in $\mathscr{M}$ and maps $\delta : U \cup \{O^*\} \to \mathbf{K}$ and $\pi : C \times U \to U(\mathscr{A})$ such that:

(a) for each $c \in C^u$ and $u \in U \cup \{O^*\}$, $\pi(c, u) \in |\mathscr{A}_{\delta(u)}|$.

(b) $\delta(O^*) = \mathbf{O}$.

(c) for each formula $\mathbf{B}_{\mathbf{x}_1 \dots \mathbf{x}_n}[u_1, \dots, u_n]$ in α where $\mathbf{B}$ contains no elements of U, we have $b(\mathfrak{A}) \models \mathbf{B}[\delta(u_1), \dots, \delta(u_n)]$.

(d) for each pair $\langle \mathbf{A}_{\mathbf{x}_1 \dots \mathbf{x}_n}[c_1, \dots, c_n] u, \rangle$ in p, we have $\mathfrak{A} \models_{\delta(u)} \mathbf{A}[\pi(c_1, u), \dots, \pi(c_n, u)]$.

We say that $\langle \delta, \pi \rangle$ *satisfies* $\langle \alpha, p \rangle$ in $\mathfrak{A}$, and that $\langle \alpha, p \rangle$ is *satisfiable in* $\mathfrak{A}$.

Define $\langle \alpha, p \rangle \leq \langle \beta, q \rangle$ iff $\alpha \subseteq \beta$ and $p \subseteq q$. Let $f(\langle \alpha, p \rangle)$ be the set of atomic sentences in α, and for $u \in U$, let $h(\langle \alpha, p \rangle, u)$ be the set of atomic sentences $\mathbf{A}$ such that $\langle \mathbf{A}, u \rangle \in p$.

In the case that Φ consists precisely of all $\Diamond\exists$-formulas, we write $\mathscr{P}(\mathscr{M})$ for the forcing condition $\mathscr{P}(\mathscr{M}, \Phi)$. If $\mathscr{M}$ is the class of all S-modal models of the S-theory T, we say that $\mathfrak{A}$ is T-*generic* iff $\mathfrak{A}$ is $\mathscr{P}(\mathscr{M})$-generic.

Now let $\langle \alpha, p \rangle$ be a condition in the forcing property $\mathscr{P}(\mathscr{M}, \Phi)$. An arbitrary S-modal structure $\mathfrak{A}$ (not necessarily an element of $\mathscr{M}$) is said to be a *model of* $\langle \alpha, p \rangle$ *at* w if there exists an interpretation $u^{\mathfrak{A}}$ in $\mathbf{K}$ of the constants u of U, together with interpretations $c^{\mathscr{A}_k}$ of the constants $c \in C^u$ in each $|\mathscr{A}_k|$ for $u^{\mathfrak{A}} = k \in \mathbf{K}$ such that each formula of α is valid in $b(\mathfrak{A})$ and for each formula $\mathbf{A}$ of p, if $k = u^{\mathfrak{A}}$, then $\mathfrak{A} \models_k \mathbf{A}$, where the constants c of $\mathbf{A}$ are interpreted as $c^{\mathscr{A}_k}$. Let $\mathbf{A} \in \Phi(c)$. We will write $\langle \alpha, p \rangle \models_u \mathbf{A}$ if every model of $\langle \alpha, p \rangle$ is model of $\mathbf{A}$ where the same interpretations of

$$\bigvee_{n<\omega} \Diamond^{t_n}\exists \mathbf{y}_1 \ldots \mathbf{y}_{i_n} \bigvee_{j_n=1}^{j_n} \mathbf{D}_{n,j}.$$

(1) **A** holds in all $\mathscr{P}(\mathscr{M}, \Phi)$-generic structures.

(2) For all $\langle u_1, \ldots, u_s\rangle \in U^s$ and all $\vec{c} \in C^m$, if $\mathfrak{A} = \langle \mathscr{A}_k, \mathbf{K}, \mathbf{R}, \mathbf{O}, \mathbf{N}\rangle$ is any $\mathscr{P}(\mathscr{M}, \Phi)$-generic model, if F_s holds in $\langle \mathbf{K}, \mathbf{R}, \mathbf{O}, \mathbf{N}\rangle$ where F_s is $\mathbf{O}^*\mathbf{R}^*u_1 \& u_1\mathbf{R}^*u_2 \wedge \ldots \wedge u_{s-1}\mathbf{R}^*u_s$, and if $k = u_s^{\mathfrak{A}}$, then $\mathfrak{A} \models_k \mathbf{B}_{\bar{\mathbf{x}}}[\vec{c}]$.

(3) For all $\langle u_1, \ldots, u_s\rangle \in U^s$ and $\vec{c} \in C^m$, if $\langle \phi, \phi\rangle \Vdash^* u_i\mathbf{R}u_{i+1}$ for $1 \le i < s$, then $\langle \phi, \phi\rangle \Vdash^*_{u_s} \mathbf{B}[c]$.

(4) For all $\langle u_1, \ldots u_s\rangle \in U^s$ and $\vec{c} \in C^m$, and for all conditions $\langle \alpha, p\rangle$ of $\mathscr{P}(\mathscr{M}, \Phi)$, if $\langle \alpha \cup \{u_i\mathbf{R}^*u_{i+1} : 1 \le i < s\}, p\rangle$ is a condition, there exist $\langle \beta, q\rangle \ge \langle \alpha, p\rangle$, an $n < \omega$, $\langle u'_1, \ldots, u'_{t_n}\rangle \in U^{t_n}$, and $\vec{d} \in C^{i_n}$ such that $u'_s\mathbf{R}^*u'_{s+1} \in \beta$ for $1 \le s < t_n$, $u_s\mathbf{R}^*u'_1 \in \beta$, and for all l with $1 \le l \le j_n$ and for all $\langle \gamma, r\rangle \ge \langle \beta, q\rangle$, there is a $\langle \delta, s\rangle \ge \langle \gamma, r\rangle$ such that $\langle \delta, s\rangle \Vdash_{u'} \mathbf{D}_{n,l}[c, \vec{d}]$, where $u' = u'_{t_n}$.

(5) For all $\langle u_1, \ldots, u_s\rangle \in U^s$ and $\vec{c} \in C^m$, and all conditions $\langle \alpha, \vec{p}\rangle$ of $\mathscr{P}(\mathscr{M}, \Phi)$, if $\langle \alpha \cup \{u_i\mathbf{R}^*u_{i+1} : 1 \le i < s\}, p\rangle$ is a condition, there exist $n < \omega$, $\langle u'_1, \ldots, u'_{t_n}\rangle \in U^{t_n}$, and $\vec{d} \in C^{i_n}$ such that

$$\langle \alpha \cup \{u_i\mathbf{R}^*u_{i+1} : 1 \le i < s\} \cup \{u_s\mathbf{R}^*u'_1\} \cup \{u'_i\mathbf{R}^*u'_{i+1} : 1 \le i < t_n\},$$
$$p \cup \{\langle \mathbf{D}_{n,l}[\vec{c}, \vec{d}], u'_{t_n}\rangle : 1 \le l \le j_n\}\rangle$$

is a condition.

(6) For all $\langle u_1, \ldots, u_s\rangle \in U^s$ and $\vec{c} \in C^m$, and all conditions $\langle \alpha, p\rangle$ of $\mathscr{P}(\mathscr{M}, \Phi)$, if $\langle \alpha \cup \{u_i\mathbf{R}^*u_{i+1} : 1 \le i < s\}, p\rangle$ is satisfiable in $\mathscr{M}$, then

$$\langle \alpha \cup \{u_i\mathbf{R}^*u_{i+1} : 1 \le i < s\}, \quad p \cup \{\langle \mathbf{B}[\vec{c}], u_s\rangle\}\rangle$$

is satisfiable in $\mathscr{M}$.

All of the equivalences except (4)$\leftrightarrow$(5) follow easily from the definitions, Lemma 13.7, and Corollary 13.14. For the equivalence (4)$\leftrightarrow$(5), we proceed as follows.

(5)$\rightarrow$(4): Let $\langle u_1, \ldots, u_s\rangle \in U^s$ and $\vec{c} \in C^m$, and let $\langle \alpha \cup \{u_i\mathbf{R}^*u_{i+1} : 1 \le i < s\}, p\rangle$ be a condition of $\mathscr{P}(\mathscr{M}, \Phi)$. Let $\langle \beta, q\rangle$ be the condition described by (5). Then $\langle \beta, q\rangle \models_u \mathbf{D}_{n,l}[\vec{c}, \vec{d}]$, where $u = u'_{t_n}$, for $1 \le l \le j_n$. Then by Lemma 14.1, $\langle \beta, q\rangle \Vdash^*_u \mathbf{D}_{n,l}[\vec{c}, \vec{d}]$ for $1 \le l \le j_n$, and this is the conclusion of (4), as desired.

(4)$\rightarrow$(5): Let $\langle u_1, \ldots, u_i\rangle \in U^s$, let $\vec{c} \in C^m$, and let $\langle \alpha \cup \{u_i\mathbf{R}u_{i+1} : 1 \le i < s\}, p\rangle$ be a condition of $\mathscr{P}(\mathscr{M}, \Phi)$. Let $\langle \beta, q\rangle$, n, $\langle u'_1, \ldots, u'_{t_n}\rangle \in U^{t_n}$, and $\vec{d} \in C^{i_n}$ be as described in (4). According to (4), for each l with $1 \le l \le j_n$, $\langle \beta, q\rangle \Vdash^*_u \mathbf{D}_{n,l}[\vec{c}, \vec{d}]$, $u = u'_{t_n}$. By our definition of conditions in

$\mathscr{P}(\mathscr{M}, \Phi)$, there must be an $\mathfrak{A} \in \mathscr{M}$ and maps δ and π such that $\langle \delta, \pi \rangle$ satisfies $\langle \beta, q \rangle$ in $\mathfrak{A}$. Let $1 \leq l \leq j_n$ and suppose that in fact $\langle \delta, \pi \rangle$ satisfied

$$\langle \beta, r \rangle = \langle \beta, q \cup \{ \langle \neg \mathbf{D}_{n,l}[\vec{c}, \vec{d}], u \rangle \} \rangle$$

in $\mathfrak{A}$. Then $\langle \beta, r \rangle$ would be a condition with $\langle \beta, r \rangle \geq \langle \beta, q \rangle$. Clearly $\langle \beta, r \rangle \models_u \neg \mathbf{D}_{n,l}[\vec{c}, \vec{d}]$, and so by Lemma 14.1,

$$\langle \beta, r \rangle \Vdash_u^* \neg \mathbf{D}_{n,l}[\vec{c}, \vec{d}].$$

Thus there is a $\langle \gamma, s \rangle \geq \langle \beta, r \rangle \geq \langle \beta, q \rangle$ such that $\langle \gamma, s \rangle \Vdash_u \neg \mathbf{D}_{n,l}[\vec{c}, \vec{d}]$ which contradicts $\langle \beta, q \rangle \Vdash_u^* \mathbf{D}_{n,l}[\vec{c}, \vec{d}]$. Thus we must have that $\langle \delta, \pi \rangle$ satisfies $\langle \beta, q \cup \{ \langle \mathbf{D}_{n,l}[\vec{c}, \vec{d}], u \rangle \} \rangle$ in $\mathfrak{A}$, and so (5) follows. ■

DEFINITION 14.4. If T is an S-theory all of whose nonlogical axioms are $\Box\forall \vee \Diamond\exists$-sentences over Φ, then T is called a *$\Box\forall \vee \Diamond\exists$-$\Phi$-theory*. If $\mathscr{M}$ is the class of all S-modal models of such a theory T, then $\mathscr{M}$ is called a *$\Box\forall \vee \Diamond\exists$-class over Φ*.

COROLLARY 14.5. If $\mathscr{M}$ is a $\Box\forall \vee \Diamond\exists$-class over Φ, then every $\mathscr{P}(\mathscr{M}, \Phi)$-generic structure belongs to $\mathscr{M}$.

Proof: Let $\mathscr{M}$ be the class of all models of T where T is a $\Box\forall \vee \Diamond\exists$-$\Phi$-theory, let $\mathbf{A} = \Box^s \forall x_1 \ldots \forall x_m \mathbf{B}$ be a nonlogical axiom of T, and let $\langle \alpha, p \rangle$ be a condition in $\mathscr{P}(\mathscr{M}, \Phi)$. Let $\mathfrak{B} \in \mathscr{M}$ be such that $\langle \alpha, p \rangle$ is satisfiable in $\mathfrak{B}$. Since $\mathfrak{B} \models \mathbf{A}$, then for any $\langle u_1, \ldots, u_s \rangle \in U^s$ and any $\vec{c} \in C^m$,

$$\langle \beta, q \rangle = \langle \alpha \cup \{ u_i \mathrm{R}^* u_{i+1} : 1 \leq i < s \}, \quad p \cup \{ \langle \mathbf{B}[\vec{c}], u_s \rangle \} \rangle$$

is satisfiable in $\mathfrak{B}$, and so $\langle \beta, q \rangle$ is satisfiable in $\mathscr{M}$. Hence by Theorem 14.3, $\mathbf{A}$ is true in all $\mathscr{P}(\mathscr{M}, \Phi)$-generic structures. Since $\mathbf{A}$ was an arbitrary nonlogical axiom of T, every $\mathscr{P}(\mathscr{M}, \Phi)$-generic structure is a model of T and hence belongs to $\mathscr{M}$. ■

Our next result partially extends Theorem 4.2 of Robinson (1970) to the present context.

THEOREM 14.5. Let T be an S-modal theory which is countable, S-consistent and possesses a model completion T*, and assume that T* is equivalent in S to a theory T′ all of whose axioms are of the form $\Box^s \forall \mathbf{x}_1 \ldots \forall \mathbf{x}_m \mathbf{A}$, where $\mathbf{A}$ is a $\Diamond\exists$-formula. Then every T-generic structure is a model of T*.

Proof: Let $\Box^s \forall \mathbf{x}_1 \ldots \forall \mathbf{x}_m \mathbf{A}$ be a nonlogical axiom of T′, let $\mathscr{M}$ be the class of all S-modal models of T, let $u_1, \ldots, u_s \in U^s$, and let $\vec{c} \in C^m$. Let α

be a finite set of basic sentences of LB[U], let $p \subseteq \Phi(C)$ be finite, where Φ is the set of $\Diamond\exists$-formulas of ML(T) and assume that the pair $\langle \alpha, p \rangle$ is satisfiable at **O** in $\mathfrak{A} \in \mathcal{M}$, say by $\langle \delta, \pi \rangle$. Since $\mathfrak{A} \in \mathcal{M}$, then $\mathfrak{A}$ is an S-modal model of T. Then there exists an extension $\mathfrak{B}$ of $\mathfrak{A}$ which belongs to $\mathcal{M}$ and is an S-modal model of T*. Since α consists of basic sentences and $p \subseteq \Phi(C)$, it follows by Lemma 9.3 that $\langle \delta, \pi \rangle$ satisfies $\langle \alpha, p \rangle$ in $\mathfrak{B}$. Assume that $\langle \delta, \pi \rangle$ satisfies $\langle \alpha \cup \{u_i R u_{i+1} : 1 \leq i < s\}, p \rangle$ at **O**. Since T′ and T* are equivalent in S, then $\vdash^{S}_{T^*} \Box^s \forall \mathbf{x}_1 \ldots \forall \mathbf{x}_m \mathbf{A}$ and so it follows that $\Box^s \forall \mathbf{x}_1 \ldots \forall \mathbf{x}_m \mathbf{A}$ is valid in $\mathfrak{B}$. Consequently, $\langle \delta, \pi \rangle$ satisfies

$$\langle \alpha \cup \{u_i R u_{i+1} : 1 \leq i < s\}, \quad (p \times \{O^*\}) \cup \{\langle \mathbf{A}_{\bar{\mathbf{x}}}[\bar{c}], u_s \rangle\} \rangle$$

in $\mathfrak{B}$. Then by Theorem 14.3, it follows that **A** holds in every T-generic structure. Since T′ and T* are equivalent in S, the result follows. ■

It is not clear how to remove the restriction on T* in Theorem 14.5. If the restriction in Theorem 9.3 that Γ be closed under conjunction could be removed, then the proof of Theorem 10.6 could be modified to show that every inductive S-theory would be equivalent in S to an S-theory all of whose nonlogical axioms were of the form $\Box^s \forall \mathbf{x}_1 \ldots \forall \mathbf{x}_m \mathbf{A}$, where **A** is a $\Diamond\exists$-formula. Since any model completion T* of T is easily seen to be inductive, then T* would automatically meet the restriction of Theorem 14.5.

§15. OMITTING TYPES AND A TWO-CARDINAL THEOREM

Our theorem on omitting types, like Keisler's, follows immediately from our foregoing work.

OMITTING TYPES THEOREM (15.1).* Let $\mathcal{M}$ be a $\Box\forall \vee \Diamond\exists$-class over Φ and let $\mathbf{A}_n = \Box^{s_n}\forall\mathbf{x}_1 \ldots \forall\mathbf{x}_{m_n}\mathbf{B}^n$ be a countable sequence of $\Box\forall \vee \Diamond\exists$-sentences over Φ. Suppose that for each n, each $\langle u_1, \ldots, u_{s_n}\rangle \in U^{s_n}$, each finite $p \subseteq \Phi(C) \times U$, and each finite set α of sentences of LB[U], if $\langle \alpha \cup \{u_i \mathrm{R}^* u_{i+1} : 1 \leq i < s_n\}, p\rangle$ is satisfiable in $\mathcal{M}$, then for each m_n-tuple $\vec{c} \in C^{m_n}$, $\langle \alpha \cup \{u_i \mathrm{R}^* u_{i+1} : 1 \leq i < s_n\}, p \cup \{\langle \mathbf{B}^n_{\vec{\mathbf{x}}}[\vec{c}], u_{s_n}\rangle\}\rangle$ is satisfiable in $\mathcal{M}$. Then $\mathcal{M}$ contains a countable structure in which each $\mathbf{A}_n$ holds.

Proof. We can assume that the fragment $\mathrm{ML}_{\mathcal{A}}$ is large enough to contain each $\mathbf{A}_n$. Let $\mathfrak{A}$ be a $\mathcal{P}(\mathcal{M}, \Phi)$-generic structure. Then $\mathfrak{A}$ is countable. By Theorem 14.3, each $\mathbf{A}_n$ holds in $\mathfrak{A}$, and by Corollary 14.5, $\mathfrak{A}$ belongs to $\mathcal{M}$. ■

The notion of direct system and limit presented in Bowen (1975) is the natural extension of the classical one and the theorems stated there for it are correct. However, it is inadequate for the applications of ultralimit constructions made earlier since it requires commutativity with respect to absolute identity, while our ultrapower constructions only guarantee commutativity with respect to the equivalence relations $\equiv_k$. Thus we had to introduce the direct limit notion of §8. However, when one is only interested in building single direct systems instead of pairs of systems with commuting 'cross-over' maps, the original notion of Bowen (1975) works quite well, and in fact seems to be the only available notion for systems containing limit points since it is not clear that the construction of §8 can be extended to more general systems. Thus in the remainder of this section, our notion of direct limit will be that of Bowen (1975).

* Related results were obtained by Mortimer (1974).

THEOREM 15.2. Let T be an S-theory in the countable language ML and suppose that ML possesses a unary predicate Z(x). Moreover, suppose that T possesses an S-modal model $\mathfrak{A} = \langle \mathscr{A}_k, \mathbf{K}, \mathbf{R}, \mathbf{O}, \mathbf{N} \rangle$ such that

$$\aleph_0 \leq \mathrm{card}(Z_{\mathscr{A}_\mathbf{O}}) < \mathrm{card}(|\mathscr{A}_\mathbf{O}|).$$

Then T possesses an S-modal model $\mathfrak{B} = \langle \mathscr{B}_l, \mathbf{L}, \mathbf{S}, \mathbf{P}, \mathbf{M} \rangle$ such that $\mathfrak{A} \equiv \mathfrak{B}$, $\mathrm{card}(Z_{\mathscr{B}_\mathbf{P}}) = \aleph_0$ and $\mathrm{card}(|\mathscr{B}_\mathbf{P}|) = \aleph_1$.

Proof. Let $\mathrm{Card}(Z_{\mathscr{A}_\mathbf{O}}) = \lambda$; a simple application of the first form of the Downward Lowenheim–Skolem Theorem (5.1) shows that we may assume that $\mathrm{card}(|\mathscr{A}_\mathbf{O}|) = \lambda^+$. Now add a new binary predicate symbol $<$ to ML. Let $\mathfrak{A}' = \langle \mathscr{A}'_k, \mathbf{K}, \mathbf{R}, \mathbf{O}, \mathbf{N} \rangle$ where $\mathscr{A}'_k = (\mathscr{A}_k, <_k)$, $<_\mathbf{O}$ well-orders $|\mathscr{A}_\mathbf{O}|$ in order type λ^+, and for $\mathbf{O}\mathbf{R}k$, if $a \in |\mathscr{A}_\mathbf{O}|$ and $b \in |\mathscr{A}_k| - |\mathscr{A}_\mathbf{O}|$, then $a <_k b$. Again using Theorem 5.1, let $\mathfrak{C} = \langle \mathscr{C}_k, \mathbf{K}, \mathbf{R}, \mathbf{O}, \mathbf{N} \rangle$ be such that $|\mathscr{C}_\mathbf{O}|$ is countable and $\mathrm{id}: \mathfrak{C} \to \mathfrak{A}'$ is an elementary embedding, $\mathfrak{C}$ being a structure for $\mathrm{ML}' = \mathrm{ML} + \{<\}$.

Let ML″ be obtained from ML′ by adding names i_a for each $a \in |\mathscr{C}_\mathbf{O}|$ and let $\mathrm{ML}^\#$ be obtained from ML″ by adding one new individual constant d. Extend $\mathfrak{C}$ to a structure for ML″ by giving the natural definitions $(i_a)_{\mathscr{C}_k} = a$. Next let T′ be the S-theory with language $\mathrm{ML}^\#$ such that the nonlogical axioms of T′ consist of all sentences of ML″ which are true in $\mathfrak{C}$ together with all sentences of the form $i_a < d$ for $a \in |\mathscr{C}_\mathbf{O}|$. Then the following is immediately clear:

(*) Let $\varphi(d)$ be a sentence. Then $\mathrm{T}' \cup \{\varphi(d)\}$ is consistent iff for arbitrarily large a in $|\mathscr{C}_\mathbf{O}|$ in the sense of $<_\mathbf{O}$, $\mathfrak{C} \models_\mathbf{O} \varphi(a)$.

For each $a \in |\mathscr{C}_\mathbf{O}|$, let $\varphi_a(x)$ be the formula $\neg x < i_a \vee \bigvee_{b \in |\mathscr{C}_\mathbf{O}|} x = i_b$, so that $\forall x \varphi_a(x)$ is equivalent to

$$\forall x [x < i_a \to \bigvee_{b \in |\mathscr{C}_\mathbf{O}|} x = i_b].$$

Now let Φ consist of all formulas of $\mathrm{ML}^\#$, so that each $\forall x \varphi_a(x)$ is a $\Box\forall \vee \Diamond\exists$-sentence over Φ. Let $\mathscr{M}$ consist of all S-modal models of T′; then $\mathscr{M}$ is $\Box\forall \vee \Diamond\exists$-class over Φ. Suppose that $p \subseteq \Phi(C) \times U$ is finite, α is a finite set of sentences of LB[U], and $\langle \alpha, p \rangle$ is satisfiable in $\mathscr{M}$. Let $c \in \mathrm{C}$ and $a \in |\mathscr{C}_\mathbf{O}|$. By (*), $\varphi_a(d)$ is consistent with T′, and so it follows that $\langle \alpha, p \cup \{\langle \varphi_a(c), 0 \rangle\} \rangle$ is satisfiable in $\mathscr{M}$. Hence by the Omitting Types Theorem, T′ has a countable model $\mathfrak{D}'$ in which each φ_a holds at the distinguished world. Clearly $\mathfrak{D}' \equiv \mathfrak{C} \equiv \mathfrak{A}$. Hence by Theorem 7.1 there exists an ultrapower $\mathfrak{D}$ of $\mathfrak{D}'$ and an elementary embedding $g: \mathfrak{C} \Rrightarrow \mathfrak{D}$. Now let $\mathfrak{D} = \langle \mathscr{D}_m, \mathbf{M}, \mathbf{T}, \mathbf{Q}, \mathbf{Y} \rangle$ and $\mathfrak{D}' = \langle \mathscr{D}'_m, \mathbf{M}', \mathbf{T}', \mathbf{Q}', \mathbf{Y}' \rangle$.

Note that since g is elementary, for $a \in |\mathscr{C}_{\mathbf{O}}|$,

$$g(a) = g((i_a)^{\mathfrak{C},\mathbf{O}}) = (i_a)^{\mathfrak{D},\mathbf{Q}} = (i_a)_{\mathscr{D}\mathbf{Q}}.$$

Moreover, the canonical embedding $e: \mathfrak{D}' \to \mathfrak{D}$ is elementary. Let $\mathfrak{F} = \langle \mathscr{F}_m, \mathbf{M}, \mathbf{T}, \mathbf{Q}, \mathbf{Y} \rangle$, where $\mathscr{F}_{\mathbf{Q}} = e[\mathscr{D}'_{\mathbf{Q}'}]$ and for $m \neq \mathbf{Q}$, $\mathscr{F}_m = \mathscr{D}_m$. It follows that $g: \mathfrak{C} \Rrightarrow \mathscr{F}$. Moreover, since each φ_a contains no modal operators,

$$\mathfrak{D}' \models_{\mathbf{Q}'} \forall x \varphi_a(x) \Leftrightarrow \mathscr{D}'_{\mathbf{Q}'} \models \forall x \varphi_a(x) \Leftrightarrow \mathfrak{F} \models_{\mathbf{Q}} \forall x \varphi_a(x).$$

Now since in $\mathfrak{A}$, $Z_{\mathscr{A}\mathbf{O}}$ has cardinality $< \lambda^+ = \text{card}(|\mathscr{A}_{\mathbf{O}}|)$, it follows that

$$\mathfrak{A} \models_{\mathbf{O}} \exists y \forall x [Z(x) \to x < y]$$

because λ^+ is regular and $<_{\mathbf{O}}$ well-orders $|\mathscr{A}_{\mathbf{O}}|$ in type λ^+ (cf. Jech (1971), pp. 11 and 17). Since $\mathfrak{C} \equiv \mathfrak{A}$, it follows that there is an $a \in |\mathscr{C}_{\mathbf{O}}|$ such that

$$\mathfrak{C} \models_{\mathbf{O}} \forall x [Z(x) \to x < i_a],$$

and so

$$\mathfrak{F} \models_{\mathbf{Q}} \forall x [Z(x) \to x < i_a].$$

Since $\forall x \varphi_a(x)$ holds in $\mathfrak{F}$ at $\mathbf{Q}$, it follows that $Z_{\mathfrak{F}_{\mathbf{Q}}} = g'' Z_{\mathscr{C}_{\mathbf{O}}}$. Thus we have that:

$$g: \mathfrak{C} \Rrightarrow \mathfrak{F}, \quad Z_{\mathfrak{F}_{\mathbf{Q}}} = g'' Z_{\mathscr{C}_{\mathbf{O}}}, \quad \text{and} \quad \text{card}(|\mathscr{F}_{\mathbf{Q}}|) = \aleph_0.$$

Now iterate this construction ω_1 times, taking direct limits at limit points as indicated below:

$$\begin{array}{ccccccccc} \mathfrak{A} & & & & & & & & \mathfrak{B} \\ \| & & & & & & & & \| \\ \mathfrak{C} = \mathfrak{F}_0 & \xrightarrow{g_{01}} & \mathfrak{F}_1 & \to \cdots \to & \mathfrak{F}_\alpha & \to & \mathfrak{F}_{\alpha+1} & \to \cdots & \varinjlim_{\alpha < \omega_1} \mathfrak{F}_\alpha. \end{array}$$

Taking the direct limit of the entire system, we obtain an S-modal structure $\mathfrak{B} = \langle \mathscr{B}_l, \mathbf{L}^{\cdot}\mathbf{S}, \mathbf{P}, \mathbf{X} \rangle$ and a map h such that:

$$h: \mathfrak{C} \Rrightarrow \mathfrak{B}, \quad Z_{\mathscr{B}_{\mathbf{P}}} = h'' Z_{\mathscr{C}_{\mathbf{O}}}, \quad \text{and} \quad \text{card}(|\mathscr{B}_{\mathbf{P}}|) = \aleph_1.$$

Then $\text{card}(Z_{\mathscr{B}_{\mathbf{P}}}) = \text{card}(Z_{\mathscr{C}_{\mathbf{O}}}) = \aleph_0$, and since $\mathfrak{C} \equiv \mathfrak{A}$, then $\mathfrak{B} \equiv \mathfrak{A}$, completing the proof. ■

APPENDIX

SEMANTIC TABLEAUX METHODS

In attempting to demonstrate the semantic completeness of any system of logic, one seeks to show that for any formula **A** of the system, either **A** is not valid (in the appropriate sense) or else there is a suitable derivation of **A**. The methods of Beth (1955, 1956), Beth and Nieland (1965), and Hintikka (1961) approach this disjunction as follows. Loosely speaking, given **A**, one attempts to 'construct or produce' a counter-model of **A** in a systematic manner. If this attempt succeeds, well and good: **A** is not valid. And if the attempt fails, the nature of the construction is such that, with a bit more work, the 'failed construction' can be converted into a derivation of **A**. Finally, one observes that if there exists any counter-model to **A** at all, the construction must succeed in producing one such, thus establishing the disjunction. Since this disjunction is exclusive, it is often expressed:

(A1) **A** is valid iff **A** is derivable.

Given a notion of a theory, this is usually extended to:

(A2) **A** is valid in all models of T
iff **A** is derivable in T.

If the system of logic extends classical logic (as do the systems we have considered), this is equivalent to:

(A3) T + $\{\neg \mathbf{A}\}$ has a model
iff **A** is not derivable in T.

Finally, taking **A** to be $\mathbf{B}\&\neg\mathbf{B}$, we obtain the Henkin form of completeness:

(A4) T has a model iff T is consistent.

Returning to (A1), the implication

(A5) if **A** is derivable, then **A** is valid

is just what we called the Consistency Lemma in §3, and is established by an induction on the complexity of the derivation of **A**. The semantic tableaux construction is concerned with the converse of (A5) which, as we indicated above, can be expressed in the form:

(A6) either **A** is not valid or **A** is derivable.

The semantic tableaux method begins by providing a systematic method for attempting to obtain a model in which **A** is false, where **A** can be any given formula. To begin by example, let **A** be the classical formula

$$\text{(A7)} \qquad p(c) \rightarrow q(c) \vee \neg p(f(c)).$$

If $\mathscr{A}$ is a classical structure in which (A7) is to be false, then $p(c)$ is to be true in $\mathscr{A}$, while $q(c) \vee \neg p(f(c))$ must be false in $\mathscr{A}$. For the latter, it must be the case that both $q(c)$ and $\neg p(f(c))$ are false in $\mathscr{A}$. And for $\neg p(f(c))$ to be false in $\mathscr{A}$ it must be that $p(f(c))$ is true in $\mathscr{A}$. Figure A1 schematically displays this reasoning. Each stage of the argument is numbered and at each stage, the formulas required to be true are presented on the left of the vertical bar, while those on the right are to be false.

0		■	$p(c) \rightarrow q(c) \vee \neg p(f(c))$
1	$p(c)$	■	$q(c) \vee \neg p(f(c))$
2	$p(c)$	■	$q(c), \neg p(f(c))$
3	$p(c), p(f(c))$	■	$q(c)$

Fig. A1.

Stage 3 presents atomic requirements on $\mathscr{A}$ which are necessary and sufficient to guarantee the falsity of (A7) in $\mathscr{A}$. Examination of these requirements shows that we can take the universe $|\mathscr{A}|$ to consist of the concrete symbol c alone, and specify that $f_{\mathscr{A}}(c)$ is c, $p_{\mathscr{A}}(c)$ holds and $q_{\mathscr{A}}(c)$ fails. More generally we could set:

$$\text{(A8)} \qquad \begin{cases} |\mathscr{A}| = \{c, f(c), f(f(c)), f(f(f(c))), \ldots\}, \\ f_{\mathscr{A}}(\mathbf{t}) = f(\mathbf{t}) \text{ for any } t \in |\mathscr{A}|, \\ p_{\mathscr{A}}(\mathbf{t}) \quad \text{iff } p(\mathbf{t}) \text{ occurs on the left of stage 3,} \\ q_{\mathscr{A}}(\mathbf{t}) \quad \text{iff } q(\mathbf{t}) \text{ occurs on the left of stage 3.} \end{cases}$$

El$^+$. Let Z be the set of formulas of the form $\exists\mathbf{vA}$ which occur in $\mathscr{T}(s)_0$, and for each $\exists\mathbf{vA}$ in Z, let $c_{\exists\mathbf{vA}}$ be a distinct new individual constant not occurring in $\mathscr{T}(s)$. Let t be the (unique) immediate successor of s under $\mathbf{T}$. Then $\mathscr{T}(t)_1 = \mathscr{T}(s)_1$ and

$$\mathscr{T}(t)_0 = \mathscr{T}(s)_0 \cup \{\mathbf{A_v}[c_{\exists\mathbf{vA}}] : \exists\mathbf{vA} \in Z\}.$$

('for each $\exists\mathbf{vA}$ in the left column, put $\mathbf{A_v}[c_{\exists\mathbf{vA}}]$ in the left column').

Er$^+$. Let Z be the set of formulas of the form $\exists\mathbf{vA}$ which occur in $\mathscr{T}(s)_1$ and let Tm_s be the set of all closed terms occurring in any formula in $\mathscr{T}(s)$. Let t be the (unique) immediate successor of s under t. Then $\mathscr{T}(t)_0 = \mathscr{T}(s)_0$ and

$$\mathscr{T}(t)_1 = \mathscr{T}(s)_1 \cup \{\mathbf{A_v}[\mathbf{b}] : \exists\mathbf{vA} \in Z \,\&\, \mathbf{b} \in \mathrm{Tm}_s\}.$$

('for each $\exists\mathbf{vA}$ in the right column and each $\mathbf{b} \in \mathrm{Tm}_s$, put $\mathbf{A_v}[\mathbf{b}]$ in the right column').

If $\mathfrak{B}$ is a branch in the tableaux system $(\mathbf{N}, \mathbf{T})$, define $\lim(\mathfrak{B})$ to be the ordered pair $(\Gamma(\mathfrak{B}), \Delta(\mathfrak{B}))$ where

$$\text{(A15)} \qquad \Gamma(\mathfrak{B}) = \bigcup_{s\in\mathfrak{B}} \mathscr{T}(s)_0,$$

and

$$\Delta(\mathfrak{B}) = \bigcup_{s\in\mathfrak{B}} \mathscr{T}(s)_1 .$$

For any $\mathbf{A}$ occurring in $\Gamma(\mathfrak{B}) \cup \Delta(\mathfrak{B})$, define

(A16) $\mathrm{ht}(\mathfrak{B}, \mathbf{A})$ = the least α such that there is an $s \in \mathfrak{B}$ of level α and $\mathbf{A} \in \mathscr{T}(s)_0 \cup \mathscr{T}(s)_1$.

Note that the following always holds:

(A17) $u\mathbf{T}v$ and $\mathbf{A} \in \mathscr{T}(u)_i$ implies $\mathbf{A} \in \mathscr{T}(v)_i$ for $i = 0, 1$.

For any branch $\mathfrak{B}$, define

(A18) $U(\mathfrak{B}) = \{\mathbf{b} : \mathbf{b}$ is a closed term occurring in $\Gamma(\mathfrak{B}) \cup \Delta(\mathfrak{B})\}$.

A branch $\mathfrak{B}$ is said to be *closed* if there is an $s \in \mathfrak{B}$ such that $\mathscr{T}(s)_0 \cap \mathscr{T}(s)_1 \neq \emptyset$. By (A17) this is equivalent to saying

$$\Gamma(\mathfrak{B}) \cap \Delta(\mathfrak{B}) \neq \emptyset.$$

A *tableaux system* is said to be *closed* if each of its branches is closed.

The remaining requirements on a classical tableaux system may now be expressed by:

(A19) for each s **N**, for each non-closed branch $\mathfrak{B}$ in $(N, \mathbf{T})$ through s and for each amenable formula **A** in $\mathscr{T}(s)$, there exist nodes $u, v, \in \mathfrak{B}$ with $s\mathbf{T}u$ such that level (v) is the successor of level (u) and $\mathscr{T}(v)$ is obtained from $\mathscr{T}(u)$ by application of the appropriate tableau rule or the Intro rule to formulas in $\mathscr{T}(u)$ which include **A**.

A formula is said to be *amenable* if its outermost logical operator is one of $\vee$, $\neg$, or $\exists$. (Recall that in principle these are the only operators apart from $\diamond$.)

(A20) if $\mathfrak{B}$ is any branch in $(\mathbf{N}, \mathbf{T})$, if $s \in \mathfrak{B}$, and if $\exists \mathbf{vA} \in \mathscr{T}(s)_1$, then for all $u \in \mathfrak{B}$ there exists a $w \in \mathfrak{B}$ with $u\mathbf{T}w$ and the (single) immediate successor of w is obtained by application of the tableau rule Er to $\exists \mathbf{vA}$.

(A21) Let *EQ* be the set of all universal closures of identity and equality axioms for the language at hand. If $s \in \mathbf{N}$ and the level of S is 0, then $EQ \subseteq \mathscr{T}(s)_0$.

If S is a system of logic (classical or modal) and (Γ, Δ) is a pair of sets of formulas, we say that (Γ, Δ) is S-*inconsistent* if there are $\mathbf{A}_1, \ldots, \mathbf{A}_m \in \Gamma$ such that:

there are $\mathbf{B}_1, \ldots, \mathbf{B}_n \in \Delta$ such that

(A22) $\qquad \vdash^{S} \mathbf{A}_1 \wedge \ldots \wedge \mathbf{A}_m \rightarrow \mathbf{B}_1 \vee \ldots \vee \mathbf{B}_n, \qquad \text{if } \Delta \neq \emptyset,$

or

(A23) $\qquad \vdash^{S} \neg(\mathbf{A}_1 \wedge \ldots \wedge \mathbf{A}_m). \qquad \text{if } \Delta = \emptyset.$

We say that (Γ, Δ) is S-*consistent* if it is not S-inconsistent.

Finally, the introduction rule (schema) is as follows:

Intro. If **A** is a formula which does not occur in $\mathscr{T}(s)$, there are two immediate successors, u and v, of s under **T**, and:

$$\mathscr{T}(u)_0 = \mathscr{T}(s)_0 \cup \{\mathbf{A}\}, \mathscr{T}(u)_1 = \mathscr{T}(s)_1,$$
$$\mathscr{T}(v)_0 = \mathscr{T}(s)_0, \mathscr{T}(v)_1 = \mathscr{T}(s) \cup \{\mathbf{A}\}.$$

LEMMA A24. Let $(\mathbf{N}, \mathbf{T}, \mathscr{T})$ be a classical tableaux system and let $s \in \mathbf{N}$. If $\mathscr{T}(s)$ is S-consistent, then for at least one of the immediate successors of s under **T**, say t, $\mathscr{T}(t)$ is S-consistent.

Proof. The proof proceeds by cases, according to the rule used. Consider two examples. Suppose the rule was Dl and that the two immediate

such that every formula of L occurs $\aleph_0$ times in this enumeration. Now we define the tableaux system in stages, as follows.

Stage 0. The unique $s \in \mathbf{N}$ of level 0 is such that $\mathscr{T}(s) = (\Gamma, \Delta)$.

Stage $n+1, n \geq 0$. A node s which has been placed in $\mathbf{N}$ by the end of stage n is said to be *active* if s has no successors under $\mathbf{T}$ at the end of stage n, and $\mathscr{T}(s)_0 \cap \mathscr{T}(s)_1 = \emptyset$. At stage $n+1$, consider each active node s already placed in $\mathbf{N}$. If formula $\mathbf{A}_n$ of (A29) occurs in $\mathscr{T}(s)$, apply the appropriate tableau rule to $\mathbf{A}_n$, placing the required nodes into $\mathbf{N}$. If $\mathbf{A}_n$ does not occur in $\mathscr{T}(s)$, apply the *Intro rule* to $\mathbf{A}_n$, again placing the required nodes into $\mathbf{N}$.

It is now a routine computation to verify that the system $(\mathbf{N}, \mathbf{T}, \mathscr{T})$ constructed at the end of $\aleph_0$ stages is indeed a classical tableaux system for (Γ, Δ).

To deal with the case of uncountable languages, we extend this construction in the most natural manner. Suppose that L is of cardinality κ. Then since $\kappa^2 = \kappa$, it is possible to enumerate the formulas of L in a sequence

(A30) $\quad \mathbf{A}_0, \mathbf{A}_1, \ldots, \mathbf{A}_\alpha, \ldots, \alpha < \kappa$

such that each formula of L occurs κ times in this sequence.

One then constructs $(\mathbf{N}, \mathbf{T}, \mathscr{T})$ by stages just as before, with the additional case:

Stage α, *α a limit ordinal.* For each sequence $h: \alpha \to \mathbf{N}$ such that the level of $h(\beta)$ is β for $\beta < \alpha$, there is to be a node $\gamma_h \in \mathrm{N}$ of level α which is the successor under $\mathbf{T}$ of all the $h(\beta)$ for $\beta < \alpha$, and such that

(A31) $\quad \mathscr{T}(\gamma_h)_i = \bigcup_{\beta < \alpha} \mathscr{T}(\beta)_i$ for $i = 0, 1$.

From (A31) it is easy to see that if $\mathscr{T}(\beta)$ is S-consistent for all $\beta < \alpha$, then $\mathscr{T}(\gamma_h)$ is S-consistent. Again, it is a routine computation that $(\mathbf{N}, \mathbf{T}, \mathscr{T})$ is now as desired. ■

THEOREM A32. If T is an S-consistent theory, there exists a classical structure which is a model of the theory

$$\mathrm{T}_{\mathrm{class}} = \{\mathbf{A}\colon \mathbf{A} \text{ is classical } \& \vdash_{\mathrm{T}} \mathbf{A}\}.$$

Proof. By hypothesis, $(\mathrm{T}, \emptyset)$ is an S-consistent pair. The result now follows by Lemmas A28 and A27. ■

Now we must extend the methods to the modal portion of our language.

DEFINITION A33. A *modal tableaux system* is a collection $\bar{\mathbf{K}}$ of classical tableaux systems together with a relation $\bar{\mathbf{R}}$, a subcollection $\bar{\mathbf{Q}}$, and a distinguished element $\bar{\mathbf{O}} \in \bar{\mathbf{K}}$. The relation $\bar{\mathbf{R}}$ may hold between nodes of various tableaux in various of the classical systems in the collection $\mathbf{R}$, but carries the restriction that if $r\bar{\mathbf{B}}s$, then s is the unique top-most node (i.e., node of level 0) of some classical tableaux system in the collection. And finally, for each classical tableaux system in $\bar{\mathbf{K}}$, it is required that if s is the topmost node of that classical system, then $r\bar{\mathbf{R}}s$ for some node of some other classical tableaux system, and that s can be obtained from r by applying one of the (appropriate) tableaux rules for $\diamond$.

The tableaux rules listed thus far concern the classical system of logic common to all of the modal systems considered in this book. Thus the same tableaux rules for the classical operators are used for all the systems. However, since it is the manner in which they treat the modal operator $\diamond$ which distinguishes the various modal systems, one is not surprised that the tableaux rules for $\diamond$ vary according to the system under consideration. Fortunately most can be fit into the following general schemata:

Schema Ml. If $s \in \bar{\mathbf{K}} - \bar{\mathbf{Q}}$, if $\diamond\mathbf{A}$ occurs in $\mathscr{T}(s)_0$ ('on the left'), and possibly some other conditions are satisfied, then there must be ('introduced') some other classical tableaux systems in $\bar{\mathbf{K}}$ at least one of which is in $\bar{\mathbf{K}} - \bar{\mathbf{Q}}$ and having topmost nodes $u, v, \ldots$, and $\mathscr{T}(u)_1, \mathscr{T}(v)_1, \ldots$ may 'initially' be $\emptyset$, and $\mathscr{T}(u)_0, \mathscr{T}(v)_0, \ldots$ contain $\mathbf{A}$ and possibly $\diamond\mathbf{A}$. For each of these (new) nodes, $s\bar{\mathbf{R}}u$, $s\bar{\mathbf{R}}v, \ldots$ holds, and some of $u, v, \ldots$ may belong to $\bar{\mathbf{Q}}$. (For some systems one considers not just a single $\diamond\mathbf{A}_i$, but a sequence $\diamond\mathbf{A}_1, \ldots, \diamond\mathbf{A}_n$.)

Schema Mr. If $\diamond\mathbf{A}$ occurs in $\mathscr{T}(s)_1$ ('on the right'), possibly some other conditions are satisfied, and the classical tableaux system containing s does not belong to $\bar{\mathbf{Q}}$, then for each u such that $s\bar{\mathbf{R}}u$ and possibly some other conditions are satisfied, $\mathbf{A}$, and possibly $\diamond\mathbf{A}$, must occur in $\mathscr{T}(u)_1$ ('are put on the right of u').

The following requirements must also be imposed:

(A34) If the classical tableaux system $(\mathbf{N}, \mathbf{T}, \mathscr{T})$ belongs to $\bar{\mathbf{K}}$, if $s, t \in \mathbf{N}$, if $s\mathbf{T}t$, and if $s\bar{\mathbf{R}}u$ for some u (somewhere in $\bar{\mathbf{K}}$), then $t\bar{\mathbf{R}}u$; and if $(\mathbf{N}, \mathbf{T}, \mathscr{T})$ belong to $\bar{\mathbf{Q}}$, then applications of the *Intro rule* are restricted so that if the formula $\mathbf{A}$ is of the form $\diamond\mathbf{B}$, s has only the successor u defined by $\mathscr{T}(u)_0 = \mathscr{T}(s)_0 \cup \{\mathbf{A}\}$, $\mathscr{T}(u)_1 = \mathscr{T}(s)_1$.

$\mathbf{K} = \mathscr{B}$ and $\mathbf{O}$ is the unique branch contributed by $\bar{\mathbf{O}}$;
$\mathscr{A}_k = \mathscr{A}(\mathfrak{B})$ for $k = \mathfrak{B} \in \mathscr{B} \subseteq \bar{\mathbf{K}}$;
$\mathbf{Q} = \{\mathfrak{B}: \mathfrak{B}$ is contributed by a $\mathscr{C} \in \bar{\mathbf{Q}}\}$;
if $k = \mathfrak{B}$ and $k' = \mathfrak{B}'$, then $k\mathbf{R}k'$ iff $\{\mathbf{A}: \Diamond\mathbf{A} \in \Delta(\mathfrak{B})\} \subseteq \Delta(\mathfrak{B}')$.

Note that since $\Gamma(\mathfrak{B}) \cup \Delta(\mathfrak{B})$ consists of all formulas of the underlying language of $\mathfrak{B}$ (including additional constants), when $k' \notin \mathbf{Q}$, the last condition is equivalent to:

$$k\mathbf{R}k' \quad \text{iff} \quad \{\mathbf{A}: \Box\mathbf{A} \in \Gamma(\mathfrak{B})\} \subseteq \Gamma(\mathfrak{B}').$$

LEMMA A41. For $\mathscr{B}$, $\mathscr{K}$, and $\mathfrak{A}(\mathscr{B})$ as above, for each $\mathfrak{B} = k \in \mathbf{K}$, and for each closed formula $\mathbf{A}$ of the language appropriate to $\mathfrak{B}$,

$$\mathfrak{A}(\mathscr{B}) \models_k \mathbf{A} \quad \text{iff} \quad \mathbf{A} \in \Gamma(\mathfrak{B}).$$

Proof. Proceeding by induction on the structure of $\mathbf{A}$, we may apply the proof of Lemma A27 to all cases except that in which $\mathbf{A}$ is $\Diamond\mathbf{E}$. In this case, note first that if $k \in \mathbf{Q}$, then by (A34), both sides of the equivalence are true. Now let $k \in \mathbf{K} - \mathbf{Q}$, and first suppose that $\Diamond\mathbf{E} \notin \Gamma(\mathfrak{B})$, so that $\Diamond\mathbf{E} \in \Delta(\mathfrak{B})$. Let $k\bar{\mathbf{R}}k' = \mathfrak{B}'$. Then by definition of $\mathbf{R}$, $\{\mathbf{A}: \Diamond\mathbf{A} \in \Delta(\mathfrak{B})\} \subseteq \Delta(\mathfrak{B}')$. Hence $\mathbf{E} \in \Delta(\mathfrak{B}')$. Then by induction, $\mathfrak{A}(\mathscr{B}) \models_{k'} \mathbf{E}$ is false, and so, since k' was arbitrary, $\mathfrak{A} \models_k \Diamond\mathbf{E}$ is false. Conversely, let $\Diamond\mathbf{E} \in \Gamma(\mathfrak{B})$, so that for some $s \in \mathfrak{B}$, $\Diamond\mathbf{E} \in \mathscr{T}(s)_0$. Then there is a $\mathscr{C}' \in \bar{\mathbf{K}}$ with top node t such that $s\bar{\mathbf{R}}t$, $\mathscr{C}' \notin \bar{\mathbf{Q}}$, and $\mathbf{E} \in \mathscr{T}(t)_0$. Let $k' = \mathfrak{B}' \in \mathbf{K}$ be the branch contributed by $\mathscr{C}'$. Now suppose that $\Diamond\mathrm{F} \in \Delta(\mathfrak{B})$. Then there must be a $u \in \mathfrak{B}$ with $s\mathbf{T}u$ and rule (schema) Mr is applied to $\Diamond\mathrm{F}$ at this node (by A14). Now by (A34), $u\bar{\mathbf{R}}t$ so that $\mathbf{F} \in \Delta(\mathfrak{B}')$. Thus it follows that $k\mathbf{R}k'$. By induction, $\mathfrak{A}(\mathscr{B}) \models_{k'} \mathbf{E}$, and thus $\mathfrak{A}(\mathscr{B}) \models_k \Diamond\mathbf{E}$, as desired. ■

LEMMA A42. Given any S-consistent pair (Γ, Δ), there exists a modal tableaux system $\mathscr{K} = (\bar{\mathbf{K}}, \bar{\mathbf{R}}, \bar{\mathbf{O}}, \bar{\mathbf{Q}})$ such that the topmost node of $\bar{\mathbf{O}}$ is (Γ, Δ).

Proof. One simply extends the reasoning in the proof of Lemma A28 so as to construct both $\bar{\mathbf{K}}$ (as well as $\bar{\mathbf{R}}$ and $\bar{\mathbf{Q}}$ and $\bar{\mathbf{O}}$) and all its (classical tableaux) elements by stages. If the cardinality of ML is $\kappa \geq \aleph_0$, fix an enumeration of the formulas of ML:

$$\mathbf{A}_0, \mathbf{A}_1, \ldots, \mathbf{A}_\alpha, \ldots, \alpha < \kappa,$$

such that each formula of ML occurs arbitrarily far out in this sequence, i.e., for all $\mathbf{A}$ in ML and all $\beta < \kappa$, there is an α with $\beta < \alpha < \kappa$ and $\mathbf{A}_\alpha$ is $\mathbf{A}$.

Stage 0. $\bar{\mathbf{K}}$ consists of one (incomplete) classical tableaux system $\bar{\mathbf{O}}$ which itself consists of one (topmost) node u with $\mathscr{T}(u) = (\Gamma, \Delta)$; $\bar{\mathbf{R}}$ and $\bar{\mathbf{Q}}$ are empty.

Stage $\gamma + 1 < \kappa$. A node s which, by the end of stage γ, has been placed into a classical tableaux system $\mathscr{C} = (\mathbf{N}, \mathbf{T}, \mathscr{T})$ which itself has been placed into $\bar{\mathbf{K}}$ by the end of stage γ is said to be *active* if $\mathscr{T}(s)_0 \cap \mathscr{T}(s)_1 = \emptyset$ and s has no $\mathbf{T}$-successors in $\mathbf{N}$ by the end of stage γ. At stage $\gamma + 1$ consider each active node. If formula $\mathbf{A}_\gamma$ occurs in $\mathscr{T}(s)$, apply the appropriate tableaux rule to $\mathbf{A}_\gamma$, placing the required new nodes into the classical tableaux $\mathscr{C}$, and, in the case of rule Ml, placing the required new tableaux system(s) into $\bar{\mathbf{K}}$ and altering $\bar{\mathbf{R}}$ and $\bar{\mathbf{Q}}$ accordingly. As new nodes are added, $\bar{\mathbf{R}}$ is altered so as to meet A34. If $\mathbf{A}_\gamma$ does not occur in $\mathscr{T}(s)$, the *Intro rule* is applied to $\mathbf{A}_\gamma$ and s, where if $\mathscr{C} \in \bar{\mathbf{Q}}$, the restriction of A34 is obeyed.

Stage $\gamma < \kappa$, *γ a limit ordinal.* For each $\mathscr{C} = (\mathbf{N}, \mathbf{T}, \mathscr{T}) \in \bar{\mathbf{K}}$, and for each $h: \gamma \to \mathbf{N}$ such that the level of $h(\delta)$ is δ for all $\delta < \gamma$, add a node γ_h to $\mathbf{N}$ of level δ which is the $\mathbf{T}$-successor of all (and only) the $h(\delta)$ for $\delta < \gamma$ and where

$$\mathscr{T}(\gamma_h)_i = \bigcup_{\delta < \gamma} (h(\delta))_i, \quad \text{for } i = 0, 1.$$

Modify $\bar{\mathbf{R}}$ in accordance with (A34).

It is now a straightforward computation to verify that $\mathscr{K}$ is as desired. ■

THEOREM A43. Given any S-consistent pair (Γ, Δ) of sets of closed formulas, there exists a structure $\mathfrak{A}$ with a world k such that $\mathfrak{A} \models_k \mathbf{A}$ holds for all $\mathbf{A} \in \Gamma$, and $\mathfrak{A} \models_k \mathbf{B}$ fails for all $\mathbf{B} \in \Delta$.

Proof. By Lemmas A42 and A39 there exists a modal tableaux system $\mathscr{K}$ containing an S-consistent branch system $\mathscr{B}$ such that the top node of $\bar{\mathbf{O}}$ (and hence $\mathbf{O}$) is (Γ, Δ). The result now follows by Lemma A41. ■

To complete this theorem, it need only be shown that the structure $\mathfrak{A}$ is an S-structure. To achieve this one attempts to require that at each stage α of the construction in Lemma A42, $\bar{\mathbf{R}}$ satisfies the appropriate conditions Γ_S as defined in §2.

Consider now some of the individual normal systems S. For normal S, $\bar{\mathbf{Q}} = \emptyset$, and the rule schemata Mr and Ml become (cf. Kripke (1963a)):

Ml. If $\mathscr{C} \in \bar{\mathbf{K}}$, s is a node in $\mathscr{C}$ and $\diamond \mathbf{A} \in \mathscr{T}(s)_0$, then there is to be a

BIBLIOGRAPHY

Aczel, P. H. G., 'Saturated Intuitionistic Theories', *Contributions to Mathematical Logic*, ed. by H. A. Schmidt *et al.*, North-Holland, Amsterdam, 1968, 1–11.

Åqvist, Lennart, *The Completeness of Some Modal Logics with Circumstantials, Subjective Conditionals, Transworld Identity, and Dispositional Predicates*, Filosvfiska Studier, Nr. 12, Uppsala University, Uppsala, 1971.

Barcan, R., *see* Marcus, R. Barcan.

Barwise, J. and Robinson, A., 'Completing Theories by Forcing', *Annals of Math. Logic* **2** (1970), 119–142.

Bell, J. and Slomsen, A., *Models and Ultraproducts: An Introduction*, second revised printing, North-Holland, Amsterdam and New York, 1971, ix + 322 pp.

Beth, E. W., 'Semantic Entailment and Formal Derivability', *Mededelingen der Koninklijke Nederlandse Akademie van Wetenschappen, Afd. Letterkunde* **18** (1955), 309–342.

Beth, E. W., 'Semantic Construction of Intuitionistic Logic', *Mededelingen der Koninklijke Nederlandse Akademie van Wetenschappen, Afd. Letterkunde* **19** (1956), 357–388.

Beth, E. W., *Formal Methods, An Introduction to Symbolic Logic*, Dordrecht and New York, 1962, 170 pp.

Beth, E. W., *The Foundation of Mathematics*, revised edition, North-Holland, Amsterdam, 1964 (paper edition, Harper & Row, New York, 1966), 741pp.

Beth, E. W. and Nieland, J. J. F., 'Semantic Construction of Lewis's Systems S4 and S5', *The Theory of Models*, ed. by J. W. Addison *et al.*, North-Holland, Amsterdam, 1965, 17–24.

Bowen, K. A., Review of Gabbay (1972b), *Mathematical Reviews* **45** (1973) Providence, R. I. p. 904.

Bowen, K. A., 'Normal Modal Model Theory,' *Journal of Philosophical Logic* **4** (1975), 97–131.

Carnap, R., *Meaning and Necessity*, Univ. of Chicago Press, Chicago, 1947. Available in paper edition.

Chang, C. C., 'On Unions of Chain of Models', *Proc. Amer. Math. Soc.* **10** (1959), 120–127.

Cresswell, M. J., *see* Hughes, G. E.

Feys, R., 'Les logiques nouvelles des modalités', *Rev. Néoscholastique de Phil.* **40** (1937), 517–553 and **41** (1938), 217–252.

Feys, R., *Modal Logics*, E. Nauwelaerts, Louvain and Paris, 1965.

Føllesdal, D., 'Interpretation of Quantifiers', in *Logic, Methodology, and Philosophy of Science III*, van Rootselaar and Staal (eds), North-Holland, Amsterdam, 1968, 271–281.

Frayne, T., Morel, A., and Scott, D., 'Reduced Direct Products', *Fund. Math.* **51** (1962), 195–228.

Gabbay, D. M., 'Craig's Interpolation Theorem for Modal Logics', *Conference in Mathematical Logic—London 1970*, Lecture Notes in Mathematics 255, Springer-Verlag, Berlin, 1972a, 111–127.

Gabbay, D. M., 'Model Theory for Intuitionistic Logic', *Zeit. f. math. Logik und Grund. d. Math.* **18** (1972b), 49–54.

Goldblatt, R. I., 'First-order Definability in Modal Logic', *JSL* **40** (1975), 35–40.

Halmos, P. R., *Naive Set Theory*, D. van Nostrand, Princeton, 1960.

Halmos, P. R., *Lectures on Boolean Algebras*, D. van Nostrand, Princeton, v + 147 pp, 1963.

Haspel, C., 'A Study of Some Interpretations of Modal and Intuitionistic Logics in the First Order Predicate Calculus', Thesis, Syracuse University, 1972.

Hintikka, K. J. J., 'Modality and Quantification', *Theoria* **27** (1961), 110–128.

Hintikka, K. J. J., 'The Modes of Modality', *Acta Philos. Fennica* (1963), *Modal and Many-Valued Logics*, 65–81.

Hintikka, K. J. J., 'Individuals, Possible Worlds, and Epistemic Logic', *Noûs* **1** (1967), 33–62.

Horn, A., 'On Sentences Which Are True of Direct Unions of Algebras', *JSL* **16** (1951), 14–21.

Hughes, G. E., and Cresswell, M. J., *An Introduction to Modal Logic*, Methuen, London, 1968. Reprinted in paperback edition, 1972.

Jech, T. J., *Lectures in Set Theory*, Lecture Notes in Mathematics, 217, Springer-Verlag, Berlin, 1971, vi + 137 pp.

Keisler, H. J., 'Forcing and the Omitting Types Theorem', *MAA Studies in Model Theory*, ed. by M. Morley, Math. Assoc. of America, 1973, 197 pp.

Kochen, S., 'Ultraproducts in the Theory of Models', *Annals of Mathematics* **74** (1962), 221–261.

Kopperman, R., *Model Theory and Its Applications*, Allyn and Bacon, Boston, 1972.

Kreisel, G. and Krivine, J. L., *Elements of Mathematical Logic (Model Theory)*, North-Holland, Amsterdam, 1967.

Kripke, S. A., 'A Completeness Theorem in Modal Logic', *JSL* **24** (1959), 1–14.

Kripke, S. A., 'Semantical Analysis of Modal Logic, I: Normal Modal Propositional Calculi', *Zeit. f. math. Logik und Grund. Math.* **9** (1963a), 113–116.

Kripke, S. A., 'Semantical Considerations on Modal Logics', *Acta Phil. Fennica* (1963b) *Modal and Many-Valued Logics*, pp. 83–94. Reprinted in Linsky (1971).

Kripke, S. A., 'Semantical Analysis of Intuitionistic Logic, I', in *Formal Systems and Recursive Functions*, ed. by J. Crossley and M. Dummet, North-Holland, Amsterdam, 1965a, 92–130.

Kripke, S. A., 'Semantical Analysis of Modal Logic II', in *The Theory of Models*, North-Holland, Amsterdam, 1965b, 206–220.

Kripke, S. A., 'Naming and Necessity', in *Semantics of Natural Language*, D. Davidson and G. Harman (eds.), D. Reidel Publ. Co., Dordrecht, 1972, 253–355.

Krivine, J. L., *see* Kreisel, G.

Kuratowski, K. and Mostowski, A., *Set Theory*, PWN and North-Holland, Warszawa and Amsterdam, 1968.

Langford, C., *see* Lewis, C.

Lemmon, E. J., 'New Foundations for Lewis Modal Systems', *JSL* **22** (1957), 176–186.

Lemmon, E. J., 'Algebraic Semantics for Modal Logics I', *JSL* **31** (1966), 46–65, and II, *JSL* **31** (1966), 191–218.

Lemmon, E. J. and Scott, D., 'Intensional Logic', Preliminary draft of initial chapters, by E. J. Lemmon, 1966.

Lewis, C. I. and Langford, C. H., *Symbolic Logic*, Dover, New York, 1932.

Linsky, L. (ed.), *Reference and Modality*, Oxford Univ. Press, Oxford, 1971, 176 pp. (paper).

Łoš, J., 'Quelques remarques, théorèmes et problèmes sur les classes définissables d'algèbres', *Mathematical Interpretation of Formal Systems*, by Th. Skolem, *et al.*, North-Holland, Amsterdam, 1955.

Łoš, J. and Suszko, R., 'On the Extending of Models IV: Infinite Sums of Models', *Fundamenta Mathematica* **44** (1957), 52–60.

Makinson, D., 'On Some Completeness Theorems in Modal Logic', *ZML* **12** (1966), 379–384.

Mal'cev, A., *The Mathematics of Algebraic Systems*, North-Holland, Amsterdam, 1971, xviii + 494 pp.

Marcus, R. Barcan, 'A Functional Calculus of First Order Based on Strict Implication', *J. Symb. Logic* **11** (1946a), 1–16.

Marcus, R. Barcan, 'The Deduction Theorem in a Functional Calculus of First Order Based on Strict Implication', *J. Symb. Logic* **11** (1946b), 115–118.

Marcus, R. Barcan, 'The Identity of Individuals in a Strict Functional Calculus of Second Order, *J. Symb. Logic* **12** (1947), 12–15.

Montague, R., 'Logical Necessity, Physical Necessity, Ethics, and Quantifiers', *Inquiry* **3** (1960), 259–269.

Morel, A., *see* Frayne, T.

Mortimer, M., 'Some Results in Modal Model Theory', *JSL* **39** (1974), 496–508.

Mostowski, A., *see* Kuratowski, K.

Osswald, H., 'Modelltheoretische Untersuchingen in der Kripke-semantik', *Archiv. f. Math. Logik und Grundlagenforschungen* **13** (1969), 3–21.

Robinson, A., *Complete Theories*, North-Holland, Amsterdam, 1956.

Robinson, A., *Introduction to Model Theory and to the Metamathematics of Algebra*, North-Holland, Amsterdam, 1965.

Robinson, A., 'Problems and Methods of Model Theory', *Aspects of Mathematical Logic*, E.I.M.E., Edizioni Cremonese, Rome, 1969.

Robinson, A., 'Forcing in Model Theory', *Ist. Nat. Alta Math., Symposia Math.* **5** (1970), 69–82.

Routley, R., 'Extensions of Makinson's Completeness Theorems in Modal Logic', *ZML* **16** (1970), 239–256.

Sacks, G. E., *Saturated Model Theory*, Benjamin, Reading, Mass., 1972.

Schütte, K., *Vollständige Systeme modaler und intuitionistischer Logik*, Springer, Berlin, 1970.

Scott, D., *see* Frayne, T. *et al.* and Lemmon, E. J. and Scott, D.

Segerberg, Krister, *An Essay in Classical Modal Logic*, 3 vols., Filosofiska Studier, Nr. 13, Uppsala.

Shoenfield, J. R., *Mathematical Logic*, Addison-Wesley, Reading, Mass., 1967.

Slomsen, A., *see* Bell, J. L.

Susko, R., *see* Łoš, J. and Suszko, R.

Tarski, A. and Vaught, R. L., 'Arithmetical Extensions of Relational Systems', *Compositio Mathematica* **13** (1957), 81–102.

Thomason, R. H., 'On the Strong Semantical Completeness of the Intuitionistic Predicate Calculus', *JSL* **33** (1968), 1–7.

Thomason, R. H., 'Some Completeness Results for Modal Predicate Calculi', *Philosophical Problems in Logic*, ed. by K. Lambert, D. Reidel Publ. Co., Dordrecht, Holland, 1970, pp. 56–76.

van Fraasen, B., 'Compactness and Löwenheim–Skolem Proofs in Modal Logic', *Logique et Analyse* **12** (1969), 167–178.

Vaught, R., *see* Tarski, A.

Wang, H., 'Logic of Many-Sorted Theories', *JSL* **17** (1952), 105–116.

Wang, H., *Logic, Computers, and Sets,* Chelsea Publ. Co., New York, 1970, reprint of *A Survey of Mathematical Logic,* Science Press and North-Holland, Peking and Amsterdam, 1962 and 1964.

Wright, G. H. von, *An Essay in Modal Logic,* North-Holland, Amsterdam, 1951.

Zeman, J. Jay, *Modal Logic,* Clarendon Press, Oxford, 1973, x + 302 pp.

INDEX OF NAMED THEOREMS

INDEX OF SYMBOLS

INDEX OF NAMES AND SUBJECTS

SYNTHESE LIBRARY

Studies in Epistemology, Logic, Methodology, and Philosophy of Science

1. J. M. Bocheński, *A Precis of Mathematical Logic.* 1959, X + 100 pp.
2. P. L. Guiraud, *Problèmes et méthodes de la statistique linguistique.* 1960, VI + 146 pp.
3. Hans Freudenthal (ed.), *The Concept and the Role of the Model in Mathematics and Natural and Social Sciences. Proceedings of a Colloquium held at Utrecht, The Netherlands, January 1960.* 1961, VI + 194 pp.
4. Evert W. Beth, *Formal Methods. An Introduction to Symbolic Logic and the Study of Effective Operations in Arithmetic and Logic.* 1962, XIV + 170 pp.
5. B. H. Kazemier and D. Vuysje (eds.), *Logic and Language. Studies Dedicated to Professor Rudolf Carnap on the Occasion of His Seventieth Birthday.* 1962, VI + 256 pp.
6. Marx W. Wartofsky (ed.), *Proceedings of the Boston Colloquium for the Philosophy of Science 1961-1962,* Boston Studies in the Philosophy of Science (ed. by Robert S. Cohen and Marx W. Wartofsky), Volume I. 1963, VIII + 212 pp.
7. A. A. Zinov'ev, *Philosophical Problems of Many-Valued Logic.* 1963, XIV + 155 pp.
8. Georges Gurvitch, *The Spectrum of Social Time.* 1964, XXVI + 152 pp.
9. Paul Lorenzen, *Formal Logic.* 1965, VIII + 123 pp.
10. Robert S. Cohen and Marx W. Wartofsky (eds.), *In Honor of Philipp Frank,* Boston Studies in the Philosophy of Science (ed. by Robert S. Cohen and Marx W. Wartofsky), Volume II. 1965, XXXIV + 475 pp.
11. Evert W. Beth, *Mathematical Thought. An Introduction to the Philosophy of Mathematics.* 1965, XII + 208 pp.
12. Evert W. Beth and Jean Piaget, *Mathematical Epistemology and Psychology.* 1966, XII + 326 pp.
13. Guido Küng, *Ontology and the Logistic Analysis of Language. An Enquiry into the Contemporary Views on Universals.* 1967, XI + 210 pp.
14. Robert S. Cohen and Marx W. Wartofsky (eds.), *Proceedings of the Boston Colloquium for the Philosophy of Science 1964-1966, in Memory of Norwood Russell Hanson,* Boston Studies in the Philosophy of Science (ed. by Robert S. Cohen and Marx W. Wartofsky), Volume III. 1967, XLIX + 489 pp.

15. C. D. Broad, *Induction, Probability, and Causation. Selected Papers.* 1968, XI + 296 pp.
16. Günther Patzig, *Aristotle's Theory of the Syllogism. A Logical-Philosophical Study of Book A of the Prior Analytics.* 1968, XVII + 215 pp.
17. Nicholas Rescher, *Topics in Philosophical Logic.* 1968, XIV + 347 pp.
18. Robert S. Cohen and Marx W. Wartofsky (eds.), *Proceedings of the Boston Colloquium for the Philosophy of Science 1966-1968,* Boston Studies in the Philosophy of Science (ed. by Robert S. Cohen and Marx W. Wartofsky), Volume IV. 1969, VIII + 537 pp.
19. Robert S. Cohen and Marx W. Wartofsky (eds.), *Proceedings of the Boston Colloquium for the Philosophy of Science 1966-1968,* Boston Studies in the Philosophy of Science (ed. by Robert S. Cohen and Marx W. Wartofsky), Volume V. 1969, VIII + 482 pp.
20. J.W. Davis, D. J. Hockney, and W. K. Wilson (eds.), *Philosophical Logic.* 1969, VIII + 277 pp.
21. D. Davidson and J. Hintikka (eds.), *Words and Objections: Essays on the Work of W. V. Quine.* 1969, VIII + 366 pp.
22. Patrick Suppes, *Studies in the Methodology and Foundations of Science. Selected Papers from 1911 to 1969.* 1969, XII + 473 pp.
23. Jaakko Hintikka, *Models for Modalities. Selected Essays.* 1969, IX + 220 pp.
24. Nicholas Rescher *et al.* (eds.), *Essays in Honor of Carl G. Hempel. A Tribute on the Occasion of His Sixty-Fifth Birthday.* 1969, VII + 272 pp.
25. P. V. Tavanec (ed.), *Problems of the Logic of Scientific Knowledge.* 1969, XII + 429 pp.
26. Marshall Swain (ed.), *Induction, Acceptance, and Rational Belief.* 1970, VII + 232 pp.
27. Robert S. Cohen and Raymond J. Seeger (eds.), *Ernst Mach: Physicist and Philosopher,* Boston Studies in the Philosophy of Science (ed. by Robert S. Cohen and Marx W. Wartofsky), Volume VI. 1970, VIII + 295 pp.
28. Jaakko Hintikka and Patrick Suppes, *Information and Inference.* 1970, X + 336 pp.
29. Karel Lambert, *Philosophical Problems in Logic. Some Recent Developments.* 1970, VII + 176 pp.
30. Rolf A. Eberle, *Nominalistic Systems.* 1970, IX + 217 pp.
31. Paul Weingartner and Gerhard Zecha (eds.), *Induction, Physics, and Ethics: Proceedings and Discussions of the 1968 Salzburg Colloquium in the Philosophy of Science.* 1970, X + 382 pp.
32. Evert W. Beth, *Aspects of Modern Logic.* 1970, XI + 176 pp.
33. Risto Hilpinen (ed.), *Deontic Logic: Introductory and Systematic Readings.* 1971, VII + 182 pp.
34. Jean-Louis Krivine, *Introduction to Axiomatic Set Theory.* 1971, VII + 98 pp.
35. Joseph D. Sneed, *The Logical Structure of Mathematical Physics.* 1971, XV + 311 pp.
36. Carl R. Kordig, *The Justification of Scientific Change.* 1971, XIV + 119 pp.
37. Milič Čapek, *Bergson and Modern Physics,* Boston Studies in the Philosophy of Science (ed. by Robert S. Cohen and Marx W. Wartofsky), Volume VII. 1971, XV + 414 pp.

38. Norwood Russell Hanson, *What I Do Not Believe, and Other Essays* (ed. by Stephen Toulmin and Harry Woolf), 1971, XII + 390 pp.
39. Roger C. Buck and Robert S. Cohen (eds.), *PSA 1970. In Memory of Rudolf Carnap,* Boston Studies in the Philosophy of Science (ed. by Robert S. Cohen and Marx W. Wartofsky), Volume VIII. 1971, LXVI + 615 pp. Also available as paperback.
40. Donald Davidson and Gilbert Harman (eds.), *Semantics of Natural Language.* 1972, X + 769 pp. Also available as paperback.
41. Yehoshua Bar-Hillel (ed.), *Pragmatics of Natural Languages.* 1971, VII + 231 pp.
42. Sören Stenlund, *Combinators, λ-Terms and Proof Theory.* 1972, 184 pp.
43. Martin Strauss, *Modern Physics and Its Philosophy. Selected Papers in the Logic, History, and Philosophy of Science.* 1972, X + 297 pp.
44. Mario Bunge, *Method, Model and Matter.* 1973, VII + 196 pp.
45. Mario Bunge, *Philosophy of Physics.* 1973, IX + 248 pp.
46. A. A. Zinov'ev, *Foundations of the Logical Theory of Scientific Knowledge (Complex Logic),* Boston Studies in the Philosophy of Science (ed. by Robert S. Cohen and Marx W. Wartofsky), Volume IX. Revised and enlarged English edition with an appendix, by G. A. Smirnov, E. A. Sidorenka, A. M. Fedina, and L. A. Bobrova. 1973, XXII + 301 pp. Also available as paperback.
47. Ladislav Tondl, *Scientific Procedures,* Boston Studies in the Philosophy of Science (ed. by Robert S. Cohen and Marx W. Wartofsky), Volume X. 1973, XII + 268 pp. Also available as paperback.
48. Norwood Russell Hanson, *Constellations and Conjectures* (ed. by Willard C. Humphreys, Jr.). 1973, X + 282 pp.
49. K. J. J. Hintikka, J. M. E. Moravcsik, and P. Suppes (eds.), *Approaches to Natural Language. Proceedings of the 1970 Stanford Workshop on Grammar and Semantics.* 1973, VIII + 526 pp. Also available as paperback.
50. Mario Bunge (ed.), *Exact Philosophy – Problems, Tools, and Goals.* 1973, X + 214 pp.
51. Radu J. Bogdan and Ilkka Niiniluoto (eds.), *Logic, Language, and Probability. A Selection of Papers Contributed to Sections IV, VI, and XI of the Fourth International Congress for Logic, Methodology, and Philosophy of Science, Bucharest, September 1971.* 1973, X + 323 pp.
52. Glenn Pearce and Patrick Maynard (eds.), *Conceptual Change.* 1973, XII + 282 pp.
53. Ilkka Niiniluoto and Raimo Tuomela, *Theoretical Concepts and Hypothetico-Inductive Inference.* 1973, VII + 264 pp.
54. Roland Fraïssé, *Course of Mathematical Logic* – Volume 1: *Relation and Logical Formula.* 1973, XVI + 186 pp. Also available as paperback.
55. Adolf Grünbaum, *Philosophical Problems of Space and Time.* Second, enlarged edition, Boston Studies in the Philosophy of Science (ed. by Robert S. Cohen and Marx W. Wartofsky), Volume XII. 1973, XXIII + 884 pp. Also available as paperback.
56. Patrick Suppes (ed.), *Space, Time, and Geometry.* 1973, XI + 424 pp.
57. Hans Kelsen, *Essays in Legal and Moral Philosophy,* selected and introduced by Ota Weinberger. 1973, XXVIII + 300 pp.
58. R. J. Seeger and Robert S. Cohen (eds.), *Philosophical Foundations of Science. Proceedings of an AAAS Program, 1969,* Boston Studies in the Philosophy of

Science (ed. by Robert S. Cohen and Marx W. Wartofsky), Volume XI. 1974, X + 545 pp. Also available as paperback.

59. Robert S. Cohen and Marx W. Wartofsky (eds.), *Logical and Epistemological Studies in Contemporary Physics*, Boston Studies in the Philosophy of Science (ed. by Robert S. Cohen and Marx W. Wartofsky), Volume XIII. 1973, VIII + 462 pp. Also available as paperback.
60. Robert S. Cohen and Marx W. Wartofsky (eds.), *Methodological and Historical Essays in the Natural and Social Sciences. Proceedings of the Boston Colloquium for the Philosophy of Science 1969-1972*, Boston Studies in the Philosophy of Science (ed. by Robert S. Cohen and Marx W. Wartofsky), Volume XIV. 1974, VIII + 405 pp. Also available as paperback.
61. Robert S. Cohen, J. J. Stachel and Marx W. Wartofsky (eds.), *For Dirk Struik. Scientific, Historical and Political Essays in Honor of Dirk J. Struik*, Boston Studies in the Philosophy of Science (ed. by Robert S. Cohen and Marx W. Wartofsky), Volume XV. 1974, XXVII + 652 pp. Also available as paperback.
62. Kazimierz Ajdukiewicz, *Pragmatic Logic*, transl. from the Polish by Olgierd Wojtasiewicz. 1974, XV + 460 pp.
63. Sören Stenlund (ed.), *Logical Theory and Semantic Analysis. Essays Dedicated to Stig Kanger on His Fiftieth Birthday*. 1974, V + 217 pp.
64. Kenneth F. Schaffner and Robert S. Cohen (eds.), *Proceedings of the 1972 Biennial Meeting, Philosophy of Science Association*, Boston Studies in the Philosophy of Science (ed. by Robert S. Cohen and Marx W. Wartofsky), Volume XX. 1974, IX + 444 pp. Also available as paperback.
65. Henry E. Kyburg, Jr., *The Logical Foundations of Statistical Inference*. 1974, IX + 421 pp.
66. Marjorie Grene, *The Understanding of Nature: Essays in the Philosophy of Biology*, Boston Studies in the Philosophy of Science (ed. by Robert S. Cohen and Marx W. Wartofsky), Volume XXIII. 1974, XII + 360 pp. Also available as paperback.
67. Jan M. Broekman, *Structuralism: Moscow, Prague, Paris*. 1974, IX + 117 pp.
68. Norman Geschwind, *Selected Papers on Language and the Brain*, Boston Studies in the Philosophy of Science (ed. by Robert S. Cohen and Marx W. Wartofsky), Volume XVI. 1974, XII + 549 pp. Also available as paperback.
69. Roland Fraïssé, *Course of Mathematical Logic* – Volume 2: *Model Theory*. 1974, XIX + 192 pp.
70. Andrzej Grzegorczyk, *An Outline of Mathematical Logic. Fundamental Results and Notions Explained with All Details*. 1974, X + 596 pp.
71. Franz von Kutschera, *Philosophy of Language*. 1975, VII + 305 pp.
72. Juha Manninen and Raimo Tuomela (eds.), *Essays on Explanation and Understanding. Studies in the Foundations of Humanities and Social Sciences*. 1976, VII + 440 pp.
73. Jaakko Hintikka (ed.), *Rudolf Carnap, Logical Empiricist. Materials and Perspectives*. 1975, LXVIII + 400 pp.
74. Milič Čapek (ed.), *The Concepts of Space and Time. Their Structure and Their Development*, Boston Studies in the Philosophy of Science (ed. by Robert S. Cohen and Marx W. Wartofsky), Volume XXII. 1976, LVI + 570 pp. Also available as paperback.

75. Jaakko Hintikka and Unto Remes, *The Method of Analysis. Its Geometrical Origin and Its General Significance,* Boston Studies in the Philosophy of Science (ed. by Robert S. Cohen and Marx W. Wartofsky), Volume XXV. 1974, XVIII + 144 pp. Also available as paperback.
76. John Emery Murdoch and Edith Dudley Sylla, *The Cultural Context of Medieval Learning. Proceedings of the First International Colloquium on Philosophy, Science, and Theology in the Middle Ages – September 1973,* Boston Studies in the Philosophy of Science (ed. by Robert S. Cohen and Marx W. Wartofsky), Volume XXVI. 1975, X + 566 pp. Also available as paperback.
77. Stefan Amsterdamski, *Between Experience and Metaphysics. Philosophical Problems of the Evolution of Science,* Boston Studies in the Philosophy of Science (ed. by Robert S. Cohen and Marx W. Wartofsky), Volume XXXV. 1975, XVIII + 193 pp. Also available as paperback.
78. Patrick Suppes (ed.), *Logic and Probability in Quantum Mechanics.* 1976, XV + 541 pp.
79. Hermann von Helmholtz: *Epistemological Writings. The Paul Hertz/Moritz Schlick Centenary Edition of 1921 with Notes and Commentary by the Editors.* (Newly translated by Malcolm F. Lowe. Edited with an Introduction and Bibliography, by Robert S. Cohen and Yehuda Elkana), Boston Studies in the Philosophy of Science (ed. by Robert S. Cohen and Marx W. Wartofsky), Volume XXXVII. 1977, XXXVIII+204 pp. Also available as paperback.
80. Joseph Agassi, *Science in Flux,* Boston Studies in the Philosophy of Science (ed. by Robert S. Cohen and Marx W. Wartofsky), Volume XXVIII. 1975, XXVI + 553 pp. Also available as paperback.
81. Sandra G. Harding (ed.), *Can Theories Be Refuted? Essays on the Duhem-Quine Thesis.* 1976, XXI + 318 pp. Also available as paperback.
82. Stefan Nowak, *Methodology of Sociological Research: General Problems.* 1977, XVIII + 504 pp.
83. Jean Piaget, Jean-Blaise Grize, Alina Szeminska, and Vinh Bang, *Epistemology and Psychology of Functions,* Studies in Genetic Epistemology, Volume XXIII. 1977, XIV+205 pp.
84. Marjorie Grene and Everett Mendelsohn (eds.), *Topics in the Philosophy of Biology,* Boston Studies in the Philosophy of Science (ed. by Robert S. Cohen and Marx W. Wartofsky), Volume XXVII. 1976, XIII + 454 pp. Also available as paperback.
85. E. Fischbein, *The Intuitive Sources of Probabilistic Thinking in Children.* 1975, XIII + 204 pp.
86. Ernest W. Adams, *The Logic of Conditionals. An Application of Probability to Deductive Logic.* 1975, XIII + 156 pp.
87. Marian Przełęcki and Ryszard Wójcicki (eds.), *Twenty-Five Years of Logical Methodology in Poland.* 1977, VIII + 803 pp.
88. J. Topolski, *The Methodology of History.* 1976, X + 673 pp.
89. A. Kasher (ed.), *Language in Focus: Foundations, Methods and Systems. Essays Dedicated to Yehoshua Bar-Hillel,* Boston Studies in the Philosophy of Science (ed. by Robert S. Cohen and Marx W. Wartofsky), Volume XLIII. 1976, XXVIII + 679 pp. Also available as paperback.
90. Jaakko Hintikka, *The Intentions of Intentionality and Other New Models for Modalities.* 1975, XVIII + 262 pp. Also available as paperback.

91. Wolfgang Stegmüller, *Collected Papers on Epistemology, Philosophy of Science and History of Philosophy,* 2 Volumes, 1977, XXVII + 525 pp.
92. Dov M. Gabbay, *Investigations in Modal and Tense Logics with Applications to Problems in Philosophy and Linguistics.* 1976, XI + 306 pp.
93. Radu J. Bogdan, *Local Induction.* 1976, XIV + 340 pp.
94. Stefan Nowak, *Understanding and Prediction: Essays in the Methodology of Social and Behavioral Theories.* 1976, XIX + 482 pp.
95. Peter Mittelstaedt, *Philosophical Problems of Modern Physics,* Boston Studies in the Philosophy of Science (ed. by Robert S. Cohen and Marx W. Wartofsky), Volume XVIII. 1976, X + 211 pp. Also available as paperback.
96. Gerald Holton and William Blanpied (eds.), *Science and Its Public: The Changing Relationship,* Boston Studies in the Philosophy of Science (ed. by Robert S. Cohen and Marx W. Wartofsky), Volume XXXIII. 1976, XXV + 289 pp. Also available as paperback.
97. Myles Brand and Douglas Walton (eds.), *Action Theory. Proceedings of the Winnipeg Conference on Human Action, Held at Winnipeg, Manitoba, Canada, 9-11 May 1975.* 1976, VI + 345 pp.
98. Risto Hilpinen, *Knowledge and Rational Belief.* 1979 (forthcoming).
99. R. S. Cohen, P. K. Feyerabend, and M. W. Wartofsky (eds.), *Essays in Memory of Imre Lakatos,* Boston Studies in the Philosophy of Science (ed. by Robert S. Cohen and Marx W. Wartofsky), Volume XXXIX. 1976, XI + 762 pp. Also available as paperback.
100. R. S. Cohen and J. J. Stachel (eds.), *Selected Papers of Léon Rosenfeld,* Boston Studies in the Philosophy of Science (ed. by Robert S. Cohen and Marx W. Wartofsky), Volume XXI. 1978, XXX + 927 pp.
101. R. S. Cohen, C. A. Hooker, A. C. Michalos, and J. W. van Evra (eds.), *PSA 1974: Proceedings of the 1974 Biennial Meeting of the Philosophy of Science Association,* Boston Studies in the Philosophy of Science (ed. by Robert S. Cohen and Marx W. Wartofsky), Volume XXXII. 1976, XIII + 734 pp. Also available as paperback.
102. Yehuda Fried and Joseph Agassi, *Paranoia: A Study in Diagnosis,* Boston Studies in the Philosophy of Science (ed. by Robert S. Cohen and Marx W. Wartofsky), Volume L. 1976, XV + 212 pp. Also available as paperback.
103. Marian Przełęcki, Klemens Szaniawski, and Ryszard Wójcicki (eds.), *Formal Methods in the Methodology of Empirical Sciences.* 1976, 455 pp.
104. John M. Vickers, *Belief and Probability.* 1976, VIII + 202 pp.
105. Kurt H. Wolff, *Surrender and Catch: Experience and Inquiry Today,* Boston Studies in the Philosophy of Science (ed. by Robert S. Cohen and Marx W. Wartofsky), Volume LI. 1976, XII + 410 pp. Also available as paperback.
106. Karel Kosík, *Dialectics of the Concrete,* Boston Studies in the Philosophy of Science (ed. by Robert S. Cohen and Marx W. Wartofsky), Volume LII. 1976, VIII + 158 pp. Also available as paperback.
107. Nelson Goodman, *The Structure of Appearance,* Boston Studies in the Philosophy of Science (ed. by Robert S. Cohen and Marx W. Wartofsky), Volume LIII. 1977, L + 285 pp.
108. Jerzy Giedymin (ed.), *Kazimierz Ajdukiewicz: The Scientific World-Perspective and Other Essays, 1931 - 1963.* 1978, LIII + 378 pp.

109. Robert L. Causey, *Unity of Science*. 1977, VIII+185 pp.
110. Richard E. Grandy, *Advanced Logic for Applications*. 1977, XIV + 168 pp.
111. Robert P. McArthur, *Tense Logic*. 1976, VII + 84 pp.
112. Lars Lindahl, *Position and Change: A Study in Law and Logic*. 1977, IX + 299 pp.
113. Raimo Tuomela, *Dispositions*. 1978, X + 450 pp.
114. Herbert A. Simon, *Models of Discovery and Other Topics in the Methods of Science*, Boston Studies in the Philosophy of Science (ed. by Robert S. Cohen and Marx W. Wartofsky), Volume LIV. 1977, XX + 456 pp. Also available as paperback.
115. Roger D. Rosenkrantz, *Inference, Method and Decision*. 1977, XVI + 262 pp. Also available as paperback.
116. Raimo Tuomela, *Human Action and Its Explanation. A Study on the Philosophical Foundations of Psychology*. 1977, XII + 426 pp.
117. Morris Lazerowitz, *The Language of Philosophy. Freud and Wittgenstein*, Boston Studies in the Philosophy of Science (ed. by Robert S. Cohen and Marx W. Wartofsky), Volume LV. 1977, XVI + 209 pp.
118. Tran Duc Thao, *Origins of Language and Consciousness*, Boston Studies in the Philosophy of Science (ed. by Robert S. Cohen and Marx. W. Wartofsky), Volume LVI. 1979 (forthcoming).
119. Jerzy Pelč, *Semiotics in Poland, 1894 - 1969*. 1977, XXVI + 504 pp.
120. Ingmar Pörn, *Action Theory and Social Science. Some Formal Models*. 1977, X + 129 pp.
121. Joseph Margolis, *Persons and Minds, The Prospects of Nonreductive Materialism*, Boston Studies in the Philosophy of Science (ed. by Robert S. Cohen and Marx W. Wartofsky), Volume LVII. 1977, XIV + 282 pp. Also available as paperback.
122. Jaakko Hintikka, Ilkka Niiniluoto, and Esa Saarinen (eds.), *Essays on Mathematical and Philosophical Logic. Proceedings of the Fourth Scandinavian Logic Symposium and of the First Soviet-Finnish Logic Conference, Jyväskylä, Finland, 1976*. 1978, VIII + 458 pp. + index.
123. Theo A. F. Kuipers, *Studies in Inductive Probability and Rational Expectation*. 1978, XII + 145 pp.
124. Esa Saarinen, Risto Hilpinen, Ilkka Niiniluoto, and Merrill Provence Hintikka (eds.), *Essays in Honour of Jaakko Hintikka on the Occasion of His Fiftieth Birthday*. 1978, IX + 378 pp. + index.
125. Gerard Radnitzky and Gunnar Andersson (eds.), *Progress and Rationality in Science*, Boston Studies in the Philosophy of Science (ed. by Robert S. Cohen and Marx W. Wartofsky), Volume LVIII. 1978, X + 400 pp. + index. Also available as paperback.
126. Peter Mittelstaedt, *Quantum Logic*. 1978, IX + 149 pp.
127. Kenneth A. Bowen, *Model Theory for Modal Logic. Kripke Models for Modal Predicate Calculi*. 1978, X + 128 pp.
128. Howard Alexander Bursen, *Dismantling the Memory Machine. A Philosophical Investigation of Machine Theories of Memory*. 1978, XIII + 157 pp.
129. Marx W. Wartofsky, *Models: Representation and Scientific Understanding*, Boston Studies in the Philosophy of Science (ed. by Robert S. Cohen and Marx W. Wartofsky), Volume XLVIII. 1979 (forthcoming). Also available as a paperback.
130. Don Ihde, *Technics and Praxis. A Philosophy of Technology*, Boston Studies in

the Philosophy of Science (ed. by Robert S. Cohen and Marx W. Wartofsky), Volume XXIV. 1979 (forthcoming). Also available as a paperback.

131. Jerzy J. Wiatr (ed.), *Polish Essays in the Methodology of the Social Sciences,* Boston Studies in the Philosophy of Science (ed. by Robert S. Cohen and Marx W. Wartofsky), Volume XXIX. 1979 (forthcoming). Also available as a paperback.
132. Wesley C. Salmon (ed.), *Hans Reichenbach: Logical Empiricist.* 1979 (forthcoming).
133. R.-P. Horstmann and L. Krüger (eds.), *Transcendental Arguments and Science.* 1979 (forthcoming). Also available as a paperback.

SYNTHESE HISTORICAL LIBRARY

Texts and Studies
in the History of Logic and Philosophy

Editors:

1. M. T. Beonio-Brocchieri Fumagalli, *The Logic of Abelard.* Translated from the Italian. 1969, IX + 101 pp.
2. Gottfried Wilhelm Leibniz, *Philosophical Papers and Letters.* A selection translated and edited, with an introduction, by Leroy E. Loemker. 1969, XII + 736 pp.
3. Ernst Mally, *Logische Schriften,* ed. by Karl Wolf and Paul Weingartner. 1971, X + 340 pp.
4. Lewis White Beck (ed.), *Proceedings of the Third International Kant Congress.* 1972, XI + 718 pp.
5. Bernard Bolzano, *Theory of Science,* ed. by Jan Berg. 1973, XV + 398 pp.
6. J. M. E. Moravcsik (ed.), *Patterns in Plato's Thought. Papers Arising Out of the 1971 West Coast Greek Philosophy Conference.* 1973, VIII + 212 pp.
7. Nabil Shehaby, *The Propositional Logic of Avicenna: A Translation from al-Shifā: al-Qiyās,* with Introduction, Commentary and Glossary. 1973, XIII + 296 pp.
8. Desmond Paul Henry, *Commentary on De Grammatico: The Historical-Logical Dimensions of a Dialogue of St. Anselm's.* 1974, IX + 345 pp.
9. John Corcoran, *Ancient Logic and Its Modern Interpretations.* 1974, X + 208 pp.
10. E. M. Barth, *The Logic of the Articles in Traditional Philosophy.* 1974, XXVII + 533 pp.
11. Jaakko Hintikka, *Knowledge and the Known. Historical Perspectives in Epistemology.* 1974, XII + 243 pp.
12. E. J. Ashworth, *Language and Logic in the Post-Medieval Period.* 1974, XIII + 304 pp.
13. Aristotle, *The Nicomachean Ethics.* Translated with Commentaries and Glossary by Hypocrates G. Apostle. 1975, XXI + 372 pp.
14. R. M. Dancy, *Sense and Contradiction: A Study in Aristotle.* 1975, XII + 184 pp.
15. Wilbur Richard Knorr, *The Evolution of the Euclidean Elements. A Study of the Theory of Incommensurable Magnitudes and Its Significance for Early Greek Geometry.* 1975, IX + 374 pp.
16. Augustine, *De Dialectica.* Translated with Introduction and Notes by B. Darrell Jackson. 1975, XI + 151 pp.

17. Arpád Szabó, *The Beginnings of Greek Mathematics.* 1979 (forthcoming).

18. Rita Guerlac, *Juan Luis Vives Against the Pseudodialecticians. A Humanist Attack on Medieval Logic.* Texts, with translation, introduction and notes. 1978, xiv + 227 pp. + index.

SYNTHESE LANGUAGE LIBRARY

Texts and Studies
in Linguistics and Philosophy

1. Henry Hiż (ed.), *Questions.* 1977, xvii + 366 pp.
2. William S. Cooper, *Foundations of Logico-Linguistics. A Unified Theory of Information, Language, and Logic.* 1978, xvi + 249 pp.
3. Avishai Margalit (ed.), *Meaning and Use. Papers Presented at the Second Jerusalem Philosophical Encounter, April 1976.* 1978 (forthcoming).
4. F. Guenthner and S. J. Schmidt (eds.), *Formal Semantics and Pragmatics for Natural Languages.* 1978, viii + 374 pp. + index.
5. Esa Saarinen (ed.), *Game-Theoretical Semantics.* 1978, xiv + 379 pp. + index.
6. F. J. Pelletier (ed.), *Mass Terms: Some Philosophical Problems.* 1978, xiv + 300 pp. + index.